T0259366

Collected Works of A.M. Turing

MECHANICAL INTELLIGENCE

Collected Works of A.M. Turing

Pure Mathematics
Edited by J.L. BRITTON

Mathematical Logic
Edited by R.O. GANDY and C.E.M. YATES

Mechanical Intelligence
Edited by D.C. INCE

Morphogenesis
Edited by P.T. SAUNDERS

NORTH-HOLLAND
AMSTERDAM · LONDON · NEW YORK · TOKYO

Collected Works of A.M. Turing

MECHANICAL INTELLIGENCE

Edited by

D.C. INCE
Open University, Milton Keynes, United Kingdom

1992
NORTH-HOLLAND
AMSTERDAM · LONDON · NEW YORK · TOKYO

ELSEVIER SCIENCE PUBLISHERS B.V.
Sara Burgerhartstraat 25
P.O. Box 211, 1000 AE Amsterdam, Netherlands

Distributors for the United States and Canada:

ELSEVIER SCIENCE PUBLISHING COMPANY INC.
655 Avenue of the Americas
New York, NY 10010, USA

ISBN: 0 444 88058 5

Library of Congress Cataloging-in-Publication Data

Turing, Alan Mathison, 1912–1954.
   Mechanical intelligence / edited by D.C. Ince.
      p.   cm. -- (Collected works of A.M. Turing)
   Includes bibliographical references and index.
   ISBN 0-444-88058-5
   1. Artificial intelligence.   I. Ince, D. (Darrel) II. Title.
  III. Series: Turing, Alan Mathison, 1912–1954. Works. 1990.
  Q335.5.T87   1992
  006.3--dc20   90-36187
                                             CIP

This book is printed on acid-free paper.

Transferred to digital printing 2005

Acknowledgement is gratefully made to the following for permission to reprint previously published articles by A.M. Turing:

Edinburgh University Press for "Intelligent Machinery", in: B. Meltzer and D. Michie (Editors), Machine Intelligence 5 (1969) 3–23.

MIT Press for "Proposal for the Development in the Mathematics Division of an Automatic Computing Engine (ACE)", in: B.E. Carpenter and R.N. Doran (Editors), A.M. Turing's ACE Report and Other Related Papers (1986) Chapter 2, 20–105;
   and for "Lecture to the London Mathematical Society on 20 February 1947", in: B.E. Carpenter and R.N. Doran (Editors), A.M. Turing's ACE Report and Other Related Papers (1986) Chapter 3, 106–124.

Oxford University Press for "Computing Machinery and Intelligence", MIND LIX (1950) 433–460.

Penguin Books Ltd. for "Solvable and Unsolvable Problems", Science News 31 (1954) 7–23.

Pitman Publishing for "Digital Computers Applied to Games", in: B.V. Bowden (Editor), Faster Than Thought (1953) Chapter 25, 286–310.

PREFACE

It is not in dispute that A.M. Turing was one of the leading figures in twentieth-century science. The fact would have been known to the general public sooner but for the Official Secrets Act, which prevented discussion of his wartime work. At all events it is now widely known that he was, to the extent that any single person can claim to have been so, the inventor of the "computer". Indeed, with the aid of Andrew Hodges's excellent biography, *A.M. Turing: the Enigma*, even non-mathematicians like myself have some idea of how his idea of a "universal machine" arose – as a sort of byproduct of a paper answering Hilbert's *Entscheidungsproblem*. However, his work in pure mathematics and mathematical logic extended considerably further; and the work of his last years, on morphogenesis in plants, is, so one understands, also of the greatest originality and of permanent importance.

I was a friend of his and found him an extraordinarily attractive companion, and I was bitterly distressed, as all his friends were, by his tragic death – also angry at the judicial system which helped to lead to it. However, this is not the place for me to write about him personally.

I am, though, also his legal executor, and in fulfilment of my duty I have organised the present edition of his works, which is intended to include all his mature scientific writing, including a substantial quantity of unpublished material. The edition will comprise four volumes, i.e.: *Pure Mathematics*, edited by Professor J.L. Britton; *Mathematical Logic*, edited by Professor R.O. Gandy and Professor C.E.M. Yates; *Mechanical Intelligence*, edited by Professor D.C. Ince; and *Morphogenesis*, edited by Professor P.T. Saunders.

My warmest thanks are due to the editors of the volumes, to the modern archivist at King's College, Cambridge, to Dr. Arjen Sevenster and Mr. Jan Kastelein at Elsevier (North-Holland), and to Dr. Einar H. Fredriksson, who did a great deal to make this edition possible.

P.N. FURBANK

ALAN MATHISON TURING – CHRONOLOGY

1912 Born 23 June in London, son of Julius Mathison Turing of the Indian Civil Service and Ethel Sara née Stoney

1926 Enters Sherborne School

1931 Enters King's College, Cambridge as mathematical scholar

1934 Graduates with distinction

1935 Is elected Fellow of King's College for dissertation on the Central Limit Theorem of Probability

1936 Goes to Princeton University where he works with Alonzo Church

1937 (January) His article "On Computable Numbers, with an Application to the Entscheidungsproblem" is published in *Proceedings of the London Mathematical Society*

Wins Procter Fellowship at Princeton

1938 Back in U.K. Attends course at the Government Code and Cypher School (G.C. & C.S.)

1939 Delivers undergraduate lecture-course in Cambridge and attends Wittgenstein's class on Foundations of Mathematics

4 September reports to G.C. & C.S. at Bletchley Park, in Buckinghamshire, where he heads work on German naval "Enigma" encoding machine

1942 Moves out of naval Enigma to become chief research consultant to G.C. & C.S.

In November sails to USA to establish liaison with American codebreakers

1943 January–March at Bell Laboratories in New York, working on speech-encypherment

1944 Seconded to the Special Communications Unit at Hanslope Park in north Buckinghamshire, where he works on his own speech-encypherment project *Delilah*

1945 With end of war is determined to design a prototype "universal machine" or "computer". In June is offered post with National Physical Laboratory at Teddington and begins work on ACE computer

1947 Severs relations with ACE project and returns to Cambridge

1948 Moves to Manchester University to work on prototype computer

1950 Publishes "Computing Machinery and Intelligence" in *Mind*

1951 Is elected FRS. Has become interested in problem of morphogenesis

1952 His article "The Chemical Basis of Morphogenesis" is published in *Philosophical Transactions of the Royal Society*

1954 Dies by his own hand in Wimslow (Cheshire) (7 June)

INTRODUCTION

Alan Turing's contribution to computer science was immense; not only in terms of depth, but also in terms of breadth. Today, his name tends to be frequently invoked in philosophical discussions about the nature of artificial intelligence. However, it is often forgotten that he was also a pioneer researcher in the areas of computer architecture and software engineering. A breadth of achievement that nobody has yet equaled in world computer science and, considering the rate at which the subject is developing, a breadth that nobody is ever again likely to achieve.

This volume contains details of his contributions to the development of computing. They range from a painstaking technical description of the architecture of the ACE computer, to broad philosophical descriptions of the nature of intelligence and the prospect of computers achieving the level of performance of humans.

What is surprising about the vast majority of these papers is that although many were written over thirty-five years ago, they still address major issues which are concerning computing researchers now. It is a measure of Turing's greatness that his work can live for so long, in a subject where research becomes out of date at a frightening speed.

We now briefly introduce the papers included in this volume.

Proposals for Development in the Mathematics Division of an Automatic Computing Engine (ACE) (1945)

The computer described in this document had its roots in a visit made by J.R. Womersley, Superintendent of the Mathematics Division at the National Physical Laboratory, to America in 1945 (HODGES 1983). Womersley, who was the first non-American to visit the American ENIAC computer, had read Turing's original work on Computability and had been impressed by seeing a Turing machine realised as electrical circuits in ENIAC.

The American work on computers had impressed Womersley so much that he invited Turing to join the National Physical Laboratory as a Scientific Officer. Turing's first task after joining the NPL was to produce a detailed design of an electronic universal machine; and it is this which is reproduced in this volume. This report was completed in 1945 and was placed before the executive committee of the NPL in March 1946. The difficulties that Womersley encountered in persuading the committee of the advantages of the proposal to build such a computer are graphically described in HODGES (1983).

Eventually £10 000 was allocated to a pilot computer and Turing started to refine the design of the computer and some programs. In 1947 the construction of the Test Assembly, a small experimental version of ACE, started. However, in October 1947, Turing left the NPL partly because of disappointment over the non-realisation of a full-scale ACE. In the end the Test Assembly was never realised. However, in 1949 work began on the construction of a Pilot Model of ACE. This was completed in 1950 and the computer entered regular use in 1952 (CAMPBELL-KELLY 1981).

The course of the project, including the political machinations which lead to the eventual departure of Turing from the NPL, is described in HODGES (1983).

Much of the report is now only of small historical interest, for example the mathematics of delay lines which occurs towards the end of the report. However, the report is of major importance for a number of reasons.

The first is Turing's insistence that the computer has a hardware system that would be as simple as possible. Turing's philosophy being that the main functionality of the ACE computer would be achieved by programming rather than by complex electronic circuitry. The trend in computer architectures since the publication of this report has been towards more and more complex hardware. However, the inevitable result of this has been the computer becoming increasingly baroque and inefficient. This has resulted in a new generation of very powerful Reduced Instruction Set Computers which, while not exactly matching Turing's Spartan hardware design, are conceptually much nearer to it than the vast majority of the computer architectures that have been designed over the last three decades.

The second remarkable feature of the proposal was the idea of modifying a stored program. The report on the ACE does not contain an implementation of conditional branching in terms of hardware, but implements this form of branching by the selective overwriting of instructions in memory. An idea which is staggeringly simple, which was not a feature of the original Turing machine, but which has been adopted in virtually every computer that has been built since.

The third feature of the proposal—and almost certainly the most important—was the idea of a hierarchy of programs. This was the first instance of a developer drawing attention to the fact that certain operations for a computer would be required time and time again, and that some facilities would be required for storing the programs (tables) that implement these operations, and for controlling the hierarchical execution of these programs. Turing's solution using BURY and UNBURY instructions is still the preferred method for controlling the execution of software.

It has been claimed that Turing's ideas represent the invention of the art of programming (HODGES 1983). This claim can be seen as something of a slight exaggeration as Conrad Zuse had, during World War II, worked out some similar ideas; indeed, a good case could be made that Babbage was the real father of programming. Nevertheless, the description of a software system as a hierarchic series of programs which communicate with each other is a unique insight which represents a major leap forward from the rather primitive programming ideas that were current in 1945 and 1946.

As is obvious from this introduction HODGES (1983) provides an excellent description of the development of ACE. Huskey's involvement in the ACE project and an excellent potted technical description of the ACE can be found in HUSKEY (1984). A description of the computer that eventually emerged from the NPL can be found in CAMPBELL-KELLY (1981). As well as providing an accessible introduction to many of Turing's ideas about programming this paper contains interesting material on the application of ACE. CAMPBELL-KELLY (1982) describes the growth of three schools of computer programming based on Cambridge University, Manchester University and the NPL. The conclusion that the paper arrives at about the NPL school of programming, which was a direct legacy of the work described in this article, is that it was the least developed and sophisticated but gave rise to numerical applications of computing which were world-class. A good analysis of the report can be found in CARPENTER and DORAN (1977), a paper which ought to be read in conjunction with the report.

Lecture to the London Mathematical Society on 20 February 1947 (1947)

This paper is of note for a number of reasons. First, it provides a good potted description of the ACE computer. Second, it is an early description of the use of subroutines, or subsidiary tables—an idea described in much more detail in his *Proposals for Development in the Mathematics Division of an Automatic Computing Engine (ACE)* (this volume). The context in the paper being that of using subroutines in order to evaluate a mathematical function. The third item of note is the brief mention of the use of machine learning techniques as a natural progression from conventional programming—a theme explained in more detail in *Intelligent Machinery* (this volume).

Intelligent Machinery (1948)

This is one of the most startling of the articles in terms of insight and prediction. In it, Turing predicts three main concepts: one of which has played a major part in computer science research, one which lies at the centre of commercial software development, and one which is a current burgeoning research area. The concept that lies at the centre of commercial software development is the subroutine; this was described in more detail in *Proposals for Development in the Mathematics Division of an Automatic Computing Engine (ACE)*. Suffice it to say, Turing, in this paper, describes one of the fundamental building blocks which enable software developers to produce large, complex systems and which allow the software developer to reuse chunks of software.

The concept that has fertilised a number of branches of computer science is the idea of computer-based theorem proving. Much of current artificial intelligence activity can be seen as the production of systems which carry out efficient deductions, based on facts culled from the environment in which a system is to be embedded. For example, expert systems are a highly successful product of the current boom in artificial intelligence. At their heart lies a database of rules which a human consultant employs in his work. The expert system takes these rules and attempts to make a deduction using a form of theorem proving.

For example, a medical expert system for diagnosing disease might contain a database of rules which represent the thought processes a doctor uses in order to diagnose a disease and the factors on which such a deduction is based; for example, the vital signs of a patient and the results of chemical tests. WINSTON (1979) provides a good introductory discussion of the role of deduction in artificial intelligence, while GALLIER (1986) provides a thorough treatment of the distance that computer science has traveled since this article was written.

Another example of automatic theorem proving is its use in formal methods of software development—the techniques heralded by Turing in *Checking a Large Routine*—where a developer wants to demonstrate that a design matches a mathematical functional specification or a program matches its design. Such proofs tend to be quite shallow but, unfortunately, generate a lot of mathematics. A major strand of research is the development of computer programs which carry out the proof process automatically. A good example of the type of program that has been developed is described in GOOD (1985).

The final prediction—one which lies at the centre of a very active research area—is that which concerns self-organising machines. Research scientists in artificial intelligence are currently attempting to discover whether it is possible to build networks—analogous to Turing's—which learn from experience. For example, such a system, say for recognising the faces of employees allowed to enter a secure building, will be trained by showing a number of pictures of each member of staff, together with a message from the trainer stating that these staff are to be allowed into the building. The system will then organise itself to accept those staff whose pictures it has been shown. Such pattern recognition has been handled poorly by current artificial intelligence systems, and researchers are attempting to produce orders of magnitude improvements using structures known as neural nets. These are self-organizing systems which attempt to mimic the neural connections in the human brain. Some early applications of neural networks are described in RUMELHART and MCCLELLAND (1986).

Checking a Large Routine (1949)

This paper describes the earliest attempt to use mathematics to specify the functionality of a computer program and to prove the properties of the computer program. Over the last three decades there have been a number of techniques used to check the correctness of a program or software subsystem against a specification, of which the most popular is testing: the execution of a program with data and checking—usually visually—that the output is what is expected. Unfortunately, testing has a number of drawbacks, the major one being that you can never guarantee correctness of a program, since testing is analogous to attempting to demonstrate a theorem by showing it holds in a number of cases.

A number of computer scientists, for example Hoare (HOARE 1969), have pointed out that programming languages have an exact semantics, and that it is possible to characterise the properties of constructs in a programming language by using relatively simple mathematics such as predicate calculus. This can be used in program proving by specifying what a program should do in terms of mathematics and using these semantics to demonstrate that the program meets this specification. It is popularly held that the notion of proving a program correct originated with Floyd (FLOYD 1967) and was considerably refined and developed by Hoare. This paper demonstrates that Turing had formulated the idea well before these researchers.

The legacy of program proving is now a great one. Software projects are becoming larger and larger, and software systems are becoming more and more complex. Experience over the last decade has shown that the major problems with software projects arise because of the nature of the specification medium used: normally natural language. Consequently, software engineers are increasingly turning to mathematics as a medium for the specification of systems. The mathematics used today for system specification tends to be richer than that used in this paper, employing set theory and logic—for a good example of a modern specification notation, see HAYES (1987)—and is merely a reflection of the fact that the systems of today are many orders of magnitude more complex than the systems of Turing's time. Nevertheless, the principles described in this paper are exactly the same. The errors of transcription which occur in this paper were first pointed out and rectified by Jones and Morris (MORRIS and JONES 1984).

Computing Machinery and Intelligence (1950)

This paper has its roots in an unpublished report that Turing wrote while at the NPL. In it, Turing attempts to describe an operational definition of intelligence by means of a guessing game in which a participant attempts to discern the sex of two other participants by means of questions. The paper briefly describes the architecture of digital computers; puts forward his own point of view: that in fifty years time computers would have a storage capacity which would enable them to play the guessing game, such that a participant would have no more than a 70% chance of making a correct identification of a computer after five minutes of questioning; and finally describes some objections to his own views, objections which are also outlined in *Intelligent Machinery*.

Turing's paper is, almost certainly, *the* fundamental paper on artificial intelligence and provides a theoretical base point from which subsequent discussion about the nature of thinking and its relationship to computation has been based. For an excellent compendium of analyses of the mind/computer problem which is generally sympathetic to Turing's viewpoint, see DENNETT (1978). Some early computer programs which attempted to replicate human behaviour and made superficial attempts at passing the Turing test are described in COLBY (1963, 1964), WEIZENBAUM (1966). An excellent description of the test applied to a computer program can be found in HAREL (1987).

Some hostile views can be found in SEARLE (1980, 1982), DREYFUS (1972). The latter being the work of a follower of Polanyi, one of the major opponents of Turing's point of view at the time of the writing of this article.

Although the debate about the relationship between thought and computational processes is at its height—occasioned by the dramatic rise of artificial intelligence of the last decade—there are as yet no clear cut answers to the hypotheses put forward by Turing in this paper. What can be said about this paper is that it is the focal point of such a high-level debate, and has practical ramifications for the present generation of artificial intelligence program developers.

Such developers produce, as their main product, the expert system. This is a program which atempts to replicate the skills of a consultant over a narrow domain. A major problem pointed out by artificial intelligence researchers, for example in PARTRIDGE (1987), is deciding on whether such an expert system is of a high enough quality; i.e. it provides the right answers, or in Turing's terms can pass muster in an imitation game, where the aim is to detect a highly skilled consultant rather than a computer. Turing's paper, although theoretical, provides a solid starting off point for research in this area.

Digital Computers Applied to Games (1953)

Games have always had a fascination for computer scientists. In particular chess has provided an ideal laboratory for the study of artificial intelligence techniques. It provides a limited environment for the exploration of topics such as planning, heuristic search and the role of knowledge. Another advantage is that it provides good quantitative measures of how successful artificial intelligence techniques are.

This paper was the first to point out a number of directions that artificial intelligence researchers pursued over the next two decades. It is a remarkable paper in that it predicts many of the developments that have occurred in artificial intelligence that have enabled chess playing programs, for example, to be capable of triumphing over the vast majority of human players.

In particular, Turing identified the use of evaluation functions: a rule which gave a numerical indication of the strength or weakness of a particular game position. He identified game playing as an excellent laboratory for research into cognitive processes. He predicted the change in the nature of chess game-playing as the game moved from the middle game to the end game. The paper predicts the course of current research on computer chess where researchers are attempting to replace brute-force algorithms for the end game in favour of algorithms which are based on a chess-players knowledge.

Currently, games programs—notably chess programs—have achieved a level of capability that few would have predicted two decades ago. The reason for this is that researchers have developed sophisticated searches, which examine a tree of moves, counter-moves, counter-counter moves etc. for a move which leads to a numerical advantage for the program. This paper sets the scene for the research which produced these searches.

There are two schools of research in artificial intelligence. The first attempts to build intelligent artefacts which do not depend on too close a consideration of what human thought processes are behind the system. The second attempts to use the computer as a laboratory for exploring hypotheses about human behavior.

This paper straddles both of these schools. It does so because Turing admits that his methods, at least applied to chess, are based on how chess players such as himself plan and evaluate moves. However, the paper does not wholly fall into the laboratory for cognitive processes school, in that a number of experiments, for example DE GROOT (1966), have shown that the very best chess players do not base their play on a search of a space of solutions based on a consideration of move, counter-move and counter-counter move, but base it on a consideration of the patterns that occur on the chess board.

Solvable and Unsolvable Problems (1954)

An algorithm is a procedure that a software developer defines in order to solve some problem. For example, the algorithm below describes the solution to the problem of producing an omelette.

```
REPEAT
   Find an egg
   Add egg to bowl
UNTIL three eggs have been added
Add some water to bowl
Mix contents of bowl
Add contents of bowl to frying pan
REPEAT
   Cook contents of frying pan
UNTIL solid and light brown in colour
```

Where the words between REPEAT and UNTIL are repeated until the condition after the word UNTIL is true. The existence of an efficient algorithm is a pre-condition for a program to be developed.

An algorithmic problem that admits to no algorithm is known as *non-computable*. If the non-computable problem is a decision problem: one whose answer is a simple yes or no, then the problem is known as *undecidable*. This popular article was intended as an attempt to explain, to a relatively unsophisticated audience, the nature of undecidable problems. The paper opens by describing a solvable problem: the sliding squares puzzle. Turing uses this example to graphically describe the fact that *undecidable* means that no systematic procedure is available for the solution of that problem. In the case of the sliding square puzzle there is a systematic procedure, albeit a computationally inefficient one, which enables the problem to be solved.

The remainder of the article concentrates on one particular undecidable problem; the word problem for groups. It was announced in 1952 that this problem was undecidable (NOVIKOV 1952) and Turing's paper attempts to provide an easily accessible introduction to the subject by giving analogical examples. The article requires close reading. Undecidability is still a very difficult subject to describe to the layperson and, although Turing does an excellent job in describing the problem, its very nature means that the prose can be difficult to understand at times. In particular page 20 onwards requires full attention from the reader.

The article describes an area of applied mathematics that has blossomed since the date of this article. A major area of research, particularly in the United States, is algorithmic complexity; i.e., the study of the questions that are posed by Turing in pages 16 and 17. A good introduction to this subject, for the relatively inexperienced, can be found in HAREL (1987) where both the word correspondence problem and the Halting problem (TURING 1937) are described.

Darrel C. INCE
February 1989

CONTENTS

2 Proposal for Development in the Mathematics Division of an Automatic Computing Engine (ACE)
A. M. Turing

Proposed Electronic Calculator

Part I. Descriptive Account

1. Introductory 2. Composition of the Calculator 3. Storages
4. Arithmetical Considerations 5. Fundamental Circuit Elements
6. Outline of Logical Control 7. External Organs 8. Scope of the
Machine 9. Checking 10. Time-Table, Cost, Nature of Work, Etc.

1. Introductory

Calculating machinery in the past has been designed to carry out accurately
and moderately quickly small parts of calculations which frequently recur.
The four processes addition, subtraction, multiplication and division, together perhaps with sorting and interpolation, cover all that could be done
until quite recently, if we except machines of the nature of the differential   [[1]]
analyser and wind tunnels, etc. which operate by measurement rather than
by calculation.

It is intended that the electronic calculator now proposed should be
different in that it will tackle whole problems. Instead of repeatedly using
human labour for taking material out of the machine and putting it back at   [[2]]
the appropriate moment all this will be looked after by the machine itself.
This arrangement has very many advantages.

(1) The speed of the machine is no longer limited by the speed of the human
operator.
(2) The human element of fallibility is eliminated, although it may to an   [[3]]
extent be replaced by mechanical fallibility.
(3) Very much more complicated processes can be carried out than could
easily be dealt with by human labour.

Once the human brake is removed the increase in speed is enormous. For
example, it is intended that multiplication of two ten figure numbers shall be
carried out in 500 μs. This is probably about 20,000 times faster than the
normal speed with calculating machines.

It is evident that if the machine is to do all that is done by the normal
human operator it must be provided with the analogues of three things, viz.
firstly, the computing paper on which the computer writes down his results
and his rough workings; secondly, the instructions as to what processes are

to be applied; these the computer will normally carry in his head; thirdly, the function tables used by the computer must be available in appropriate form to the machine. These requirements all involve *storage of information* or *mechanical memory*. This is not the place for a detailed discussion of the various kinds of storage available* and the considerations which govern their usefulness and which limit what we can expect. For the present let us only remark that the memory needs to be very large indeed by comparison with standards which prevail in most valve and relay work, and that it is necessary therefore to look for some more economical form of storage.

〚4〛   It is intended that the setting up of the machine for new problems shall be virtually only a matter of paper work. Besides the paper work nothing will have to be done except to prepare a pack of Hollerith cards in accordance with this paper work, and to pass them through a card reader connected with the machine. There will positively be no internal alterations to be made even if we wish suddenly to switch from calculating the energy levels of the neon atom to the enumeration of groups of order 720. It may appear somewhat puzzling that this can be done. How can one expect a machine to do all this multitudinous variety of things? The answer is that we should consider the machine as doing something quite simple, namely carrying out orders given to it in a standard form which it is able to understand.

The actual calculation done by the machine will be carried out in the binary scale. Material will however be put in and taken out in decimal form.

In order to obtain high speeds of calculation the calculator will be entirely electronic. A unit operation (typified by adding one and one) will take 1 microsecond. It is not thought wise to design for higher speeds than this as yet.

The present report gives a fairly complete account of the proposed calculator. It is recommended however that it be read in conjunction with J. von Neumann's 'Report on the EDVAC'.

2. Composition of the Calculator

We list here the main components of the calculator as at present conceived:

〚5〛   (1) Erasable memory units of fairly large capacity, to be known as dynamic storage (DS). Probably consisting of between 50 and 500 mercury tanks with a capacity of about 1000 digits each.

*See §16.

〚2〛

(2) Quick reference temporary storage units (TS) probably numbering about 50 and each with a capacity of say 32 binary digits.

(3) Input organ (IO) to transfer instructions and other material into the calculator from the outside world. It will have a mechanical part consisting of a Hollerith card reading unit, and an electronic part which will be internal to the calculator.

(4) Output organ (OO), to transfer results out of the calculator. It will have an external part consisting of a Hollerith card reproducer and an internal electronic part.

(5) The logical control (LC). This is the very heart of the machine. Its purpose is to interpret the instructions and give them effect. To a large extent it merely passes the instructions on to CA. There is no very distinct line between LC and CA.

(6) The central arithmetic part (CA). If we like to consider LC as the analogue of a computer then CA must be considered a desk calculating machine. It carries out the four fundamental arithmetical processes (with possible exception of division, see p. 68), and various others of the nature of copying, substituting, and the like. To a large extent these processes can be reduced to one another by various roundabout means; judgment is there-   〚6〛
fore required in choosing an appropriate set of fundamental processes.

(7) Various 'trees' required in connection with LC and CA for the selection of the information required at any moment. These trees require much more valve equipment than LC and CA themselves.

(8) The clock (CL). This provides pulses, probably at a recurrence frequency of a megacycle, which are applied, together with gating signals, to the grids of most of the valves. It provides the synchronisation for the whole calculator.

(9) Temperature control system for the delay lines. This is a somewhat mundane matter, but is important.

(10) Binary to decimal and decimal to binary converters. These will have virtually no outward and visible form. They are mentioned here lest it be thought they have been forgotten.

(11) Starting device.

(12) Power supply.

3. Storages

(i) *The storage problem.* As was explained in §1 it is necessary for the calculator to have a memory or information storage. Actually this appears to be the main limitation in the design of a calculator, i.e. if the storage problem can be solved all the rest is comparatively straightforward. In the past it has not been possible to store very large quantities of information economically in such a way that the information is readily accessible. There were economical methods such as storage on five-unit tape, but with these the information was not readily accessible, especially if one wishes to jump from point to point. There were also forms with good accessibility, such as storage on relays and valves, but those were quite prohibitively uneconomical. There are now several possibilities for combining economy with accessibility which have been developed, or are being developed. In this section we describe the one which will most probably be used in the calculator.

⟦7⟧

(ii) *Delay line storage*. All forms of storage depend on modifying in some way the physical state of some storage medium. In the case of 'delay line storage' the medium consists of mercury, water, or some other liquid in a tube or tank, and we modify its state of compression at various points along the tube. This is done by forcing supersonic waves into the tube from one end. The state of the storage medium is not constant as it would be for instance if the storage medium were paper or magnetic tape. The information moves along the tube with the speed of sound. Unless we take some precautions the sound carrying the information will pass out of the end of the tube and be lost. We can effectively prevent this by detecting the sound in some way (some form of microphone) as it comes out, and amplifying it and putting it back at the beginning. The amplifying device must correct for the attenuation of the tube, and must also correct for any distortion of form caused by the transmission through the tube, otherwise after many passages through the tube the form will be eventually completely lost. We can only restore the form of the signal satisfactorily if the various possible ideal signal forms are quite distinct, for otherwise it will not be possible to distinguish between the undistorted form of one signal and a distorted form of another. The scheme actually proposed only recognizes 2^{1024} distinct states of compression of the water medium, these being sequences of 1024 pulses of two different sizes, one of which will probably be zero. The amplifier at the end of the line always reshapes the signal to bring it back to the nearest ideal signal.

Alternatively we may consider the delay line simply as providing a delay, as its name implies. We may put a signal into the line, and it is returned to us after a certain definite delay. If we wish to make use of the information

⟦4⟧

contained in it when it comes back after being delayed we do so. Otherwise we just delay it again, and repeat until we do require it. This aspect loses sight of the fact that there is still a storage medium of some kind, with a variety of states according to the information stored.

There are, of course, other forms of delay line than those using acoustic waves.

(iii) *Technical proposals for delay line.* Let us now be more specific. It is proposed to build 'delay line' units consisting of mercury or water tubes about 5′ long and 1″ diameter in contact with a quartz crystal at each end. The velocity of sound in either mercury or water is such that the delay will be 1.024 ms. The information to be stored may be considered to be a sequence of 1024 'digits' (0 or 1), or 'modulation elements' (mark or space). These digits will be represented by a corresponding sequence of pulses. The digit 0 (or space) will be represented by the absence of a pulse at the appropriate time, the digit 1 (or mark) by its presence. This series of pulses is impressed on the end of the line by one piezo-crystal, it is transmitted down the line in the form of supersonic waves, and is reconverted into a varying voltage by the crystal at the far end. This voltage is amplified sufficiently to give an output of the order of 10 volts peak to peak and is used to gate a standard pulse generated by the clock. This pulse may be again fed into the line by means of the transmitting crystal, or we may feed in some altogether different signal. We also have the possibility of leading the gated pulse to some other part of the calculator, if we have need of that information at the time. Making use of the information does not of course preclude keeping it also. The figures above imply of course that the interval between digits is 1 μs.

It is probable that the pulse will be sent down the line as modulation on a carrier, possibly at a frequency of 15 Mc/s.

(iv) *Effects of temperature variations.* The temperature coefficient of the velocity of sound in mercury is quite small at high frequencies. If we keep the temperatures of the tanks correct to within one degree Fahrenheit it will be sufficient. It is only necessary to keep the tanks nearly at equal temperatures. We do not need to keep them all at a definite temperature: variations in the temperature of the room as a whole may be corrected by altering the clock frequency.

4. Arithmetical Considerations

(i) *Minor cycles.* It is intended to divide the information in the storages up into units, probably of 32 digits or thereabouts. Such a storage will be appropriate for carrying a single real number as a binary decimal or for carrying a single instruction. Each sub-storage of this kind is called a *minor cycle* or *word*. The longer storages of length about 1000 digits are called *major cycles*. It will be assumed for definiteness that the length of the minor cycle is 32 and that of the major 1024, although these need not yet be fixed.

(ii) *Use of the binary scale.* The binary scale seems particularly well suited for electronic computation because of its simplicity and the fact that valve equipment can very easily produce and distinguish two sizes of pulse. Apart from the input and output the binary scale will be used throughout in the calculator.

(iii) *Requirements for an arithmetical code.* Besides providing a sequence of digits the statement of the value of a real number has to do several other things. All included (probably), we must:

(a) State the digits themselves, or in other words we must specify an integer in binary form.
(b) We must specify the position of the decimal point.
(c) We must specify the sign.
(d) It would be desirable to give limits of accuracy.
(e) It would be desirable to have some reference describing the significance of the number. This reference might at the same time distinguish between [[8]] minor cycles which contain numbers and those which contain orders or other information.

None of these except for the first could be said to be absolutely indispensable, but, for instance, it would certainly be inconvenient to manage without a sign reference. The digit requirements for these various purposes are roughly:

(a) 9 decimal digits, i.e. 30 binary,
(b) 9 digits,
(c) 1 digit,
(d) 10 digits,
(e) very flexible.

(iv) *A possible arithmetical code.* It is convenient to put the digits into one minor cycle and the fussy bits into another. This may perhaps be qualified as far as the sign digit is concerned: by a trick it can be made part of the normal digit series, essentially in the same way as we regard an initial series of figures 9 as indicating a negative number in normal computing. Let

[[6]]

us now specify the code without further beating about the bush. We will use two minor cycles whose digits will be called $i_1 \ldots i_{32}, j_1 \ldots j_{32}$. Of these $j_{24} \ldots j_{32}$ are available for identification purposes, and the remaining digits make the following statement about the number ξ.

There exist rational numbers β, γ and an integer m such that

$$|\xi - 2^m \beta| < \gamma$$

$$\beta = \sum_{s=1}^{31} 2^{s-1} i_s - 2^{31} i_{32}$$

$$m = \sum_{t=1}^{9} 2^{t-1} j_t - 256$$

$$\gamma = \sum_{u=10}^{17} 2^{u+m-n} j_u$$

$$n = \sum_{v=18}^{23} 2^{v-18} j_v$$

This code allows us to specify numbers from ones which are smaller than 10^{-70} to ones which are larger than 10^{86}, mentioning a value with sufficient figures that a difference of 1 in the last place corresponds to from 2.5 to 5 parts in 10^{10}. An error can be described smaller than a unit in the last place or as large as 30,000 times the quantity itself (or by more if this quantity has its first few 'significant' digits zero).

(v) *The operations of CA.* The division of the storage into minor cycles is only of value so long as we can conveniently divide the operations to be done into unit operations to be performed on whole minor cycles. When we wish to do more elaborate types of process in which the digits get individual treatment we may find this form of division rather awkward, but we shall still be able to carry these processes out in some roundabout way provided the CA operations are sufficiently inclusive. A list is given below of the operations which will be included. Actually this account is distinctly simplified, and an accurate picture can only be obtained by reading §12. The account is however quite adequate for an understanding of the main problems involved. The list is certainly theoretically adequate, i.e. given time and instruction tables any required operation can be carried out. The operations are:

(1) Transfers of material between different temporary storages, and   [9] between temporary storages and dynamic storage.

(2) Transfers of material from the DS to cards and from cards to DS.

(3) The various arithmetical operations, addition, subtraction, and multiplication (division being omitted), also 'short multiplication' by numbers less than 16, which will be much quicker than long multiplication.

(4) To perform the various logical operations digit by digit. It will be sufficient to be able to do 'and', 'or', 'not', 'if and only if', 'never' (in symbols $A \& B, A \vee B, \sim A, A \equiv B, F$). In other words we arrange to do the processes corresponding to $xy, x + y + xy, 1 + x, 1 + (x + y)^2, 0$ digit by digit, modulo 2, where x and y are two corresponding digits from two particular TS (actually TS 9 and TS 10).

5.   Fundamental Circuit Elements

The electronic part of the calculator will be somewhat elaborate, and it will certainly not be feasible to consider the influence of every component on every other. We shall avoid the necessity of doing this if we can arrange that each component only has an appreciable influence on a comparatively small number of others. Ideally we would like to be able to consider the circuit as built up from a number of circuit elements, each of which has an output which depends only on its inputs, and not at all on the circuit into which it is working. Besides this we would probably like the output to depend only on certain special characteristics of the inputs. In addition we would often be glad for the output to appear simultaneously with the inputs.

These requirements can usually be satisfied, to a fairly high accuracy, with electronic equipment working at comparatively low frequencies. At megacycle frequencies however various difficulties tend to arise. The input capacities of valves prevent us from ignoring the nature of the circuit into which we are working; limiting circuits do not work very satisfactorily:

[[10]]   capacities and transit times are bound to cause delays between input and output. These difficulties may be best resolved by bending before the storm. The delays may be tolerated by accepting them and working out a time table which takes them into account. Indefiniteness in output may be tolerated by thinking in terms of 'classes of outputs'. Thus instead of saying 'The inputs A and B give rise to the output C', we shall say 'Inputs belonging to classes P and Q give rise to an output in class R'. The various classes must be quite distinct and must be far from overlapping, i.e. topologically

[[11]]   speaking we might say that they must be a finite distance apart. If we do this we shall have made a very definite division of labour between the mathe-

[[8]]

maticians and the engineers, which will enable both parties to carry on without serious doubts as to whether their assumptions are in agreement with those of the other party.

For the present we shall merely ignore the difficulties because we wish to illustrate the principles. We shall assume the circuit elements to have all the most agreeable properties. It may be added that this will only affect our circuits in so far as we assume instantaneous response, and that not very seriously. The questions of stable output only involve the mathematician to the extent of a few definitions.

In the present section we shall only be concerned with what the circuit elements do. A discussion of how these effects can be obtained will be given in §15. The circuit elements will be divided into valve-elements and delay elements.

(i) *Delay line, with amplifier and clock gate.* This is shown as a rectangle with an input and output lead

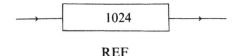

REF

the arrow at the input end faces towards the rectangle and at the output end faces away. The name of the delay line, if any, will be written outside and the delay in pulse periods inside.

This circuit element delays the input by the appropriate number of pulse periods and also standardizes it, i.e. converts it into the nearest standard form by correcting amplitude shape and time.

(ii) *The unit delay.* This is represented by a triangle, thought of as a modified form of arrow

The input to output direction is indicated by the arrow. This delay element ideally provides a delay of one pulse period.

(iii) *Limiting amplifier.* Ideally this valve-element is intended to give no output for inputs of less than a certain standard value, and to give a standard pulse as output when the input exceeds a second standard value. Intermediate input values are supposed not to occur. If we combine this with a resistance network in which a number of input signals are combined the condition takes the form that if the input signals are $s_1 s_2 \ldots s_n$ there will be zero output unless $\alpha_1 s_1 + \cdots + \alpha_n s_n \geqslant \beta_1$ and a standard or unit output if $\alpha_1 s_1 + \cdots + \alpha_n s_n > \beta_2$. This may be simplified by assuming that the inputs

$s_1 \ldots s_n$ are always either 0 or 1 and the coefficients $\alpha_1 \ldots \alpha_n$ either 1 or $-\infty$ and also by requiring the integral parts of $\beta_1 \beta_2$ to be the same. We represent the valve element by a circle, and the inputs with a line and an arrow facing towards it, the outputs with lines and arrows facing away. (Fig. 1). A coefficient $-\infty$ (*inhibitory* coupling) is shown with a small circle cutting a large circle (Fig. 2). The smallest total for which an output is obtained (i.e.

[[12]] integral part of β_1 or β_2 plus 1) is shown inside the circle, but is omitted if it is 1. This number we may call the threshold.

When we require coefficients α larger than 1 we may show more than one connection from one source. Negative coefficients may effectively be shown by means of the negation circuit ⟶⟶ which interchanges 0 and 1. Thus in the circuit of Fig. 3 the valve element D will be stimulated (i.e. emit a standard pulse) if either A is stimulated or both B and C are not.

(iv) *Trigger circuits.* A trigger circuit, which is shown as an ellipse, differs from a limiting amplifier circuit in that once the inputs have reached the threshold so that it emits one pulse, it will continue to emit pulses until it receives an inhibitory stimulus. It is in fact equivalent to a limiting amplifier with a number of excitatory connections from itself with a delay of one

[[13]] unit. Thus for instance the two circuits shown in Fig. 4 are equivalent. We show the trigger circuits with a different notation partly to simplify the drawing and partly because they will in fact be made up from different circuits. There is also another practical difference. The output from a trigger circuit will be a D.C. voltage, so long as it is not disturbed one way or the other, whereas the output from a limiting amplifier with feedback is more or less pulsiform.

(v) *Differentiator circuit and change circuit.* We sometimes wish to indicate an output from a trigger circuit either at the beginning or the end of its stimulation. This would in fact be done with a capacity resistance 'differentiator' circuit. Such a circuit designed to produce a positive (excitatory) pulse at the beginning will be denoted by ⟶(B)⟶ and one at the end by ⟶(E)⟶. These are understood to be respectively equivalent to the two circuits of Fig. 5. We may also occasionally wish to make connection to a trigger circuit in such a way that stimulus always changes the condition of the trigger circuit, either from stimulation to non-stimulation or vice-versa. This is indicated by a small square at the connection point thus

and is equivalent to Fig. 6.

[[10]]

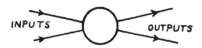

FIG. 1

FIG. 2

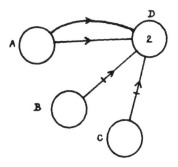

FIG 3

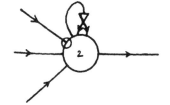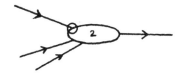

FIG. 4

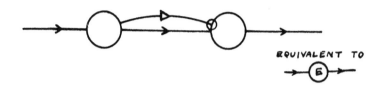

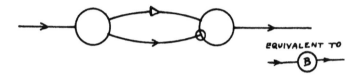

FIG. 5

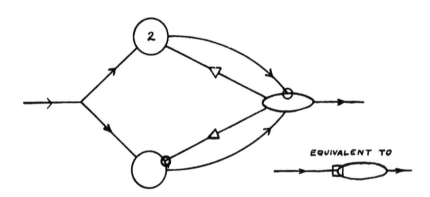

FIG. 6

[[12]]

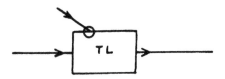

FIG. 7

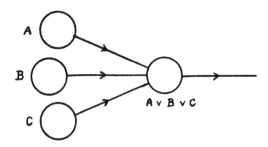

FIG. 8

(vi) *The trigger limiter*. Sometimes we wish a continuously varying voltage to initiate a train of pulses, the pulses to be synchronous with the clock and to start approximately when the continuous voltage reaches a certain value. All of the pulses that occur must be of the standard or unit size. There must definitely be no half-size pulses possible. The train of pulses may be stopped by pulses from some other source.

This valve element is indicated by a somewhat squat rectangle containing the letters TL. The continuous voltage input is shown as in an excitatory connection and the stopping pulse as an inhibitory connection, as in Fig. 7.

(vii) *The adder and other examples*. We may now illustrate the use of these circuit elements by means of some simple examples.

The simplest circuit perhaps is that for the logical 'or' (cf. p. 49). In the circuit of Fig. 8 there is an output pulse from the unnamed element if there is one from any one of A, B, C. We shall find it convenient in such cases to describe this element as $A \vee B \vee C$. The circuits of Fig. 9 are self explanatory in view of our treatment of $A \vee B \vee C$.

An adder network is shown in Fig. 10. It will add two numbers which enter along the leads shown on the left in binary form, with the least

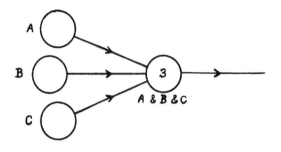

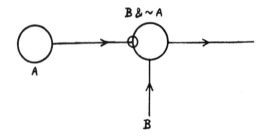

FIG. 9

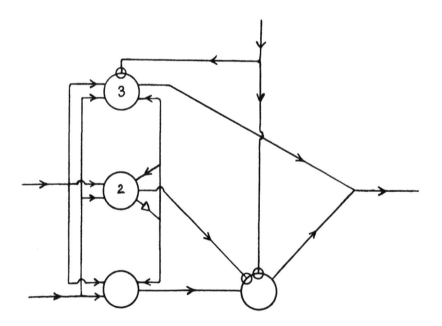

FIG. 10   ADDER NETWORK

[[14]]

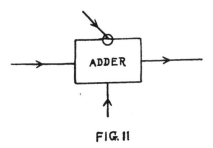

FIG. 11

significant digit first, the output appearing on the right. An input signal
from the top will inhibit any output. The method of operation is as follows.
The three valve elements on the left all have stimulation from the same
three sources, viz. the two inputs and one corresponding to the carry digit
from the last figure, which was formed by the element with threshold 2. We   [[14]]
can distinguish the four different possible totals 0, 1, 2, 3 according to
which of the valve elements are stimulated. We wish to get an output pulse
if the total is 1 or 3. This may be expressed as a pulse if the total is 3 or if it
is 1 and not 2 or more. If we write T_n to mean 'the total is n or more' the
condition is $T_3 \vee (T_1 \ \& \sim T_2)$. Using our standard networks for $A \vee B$ and
for $A \ \& \sim B$ and observing that the three valve elements on the left of the
adder are stimulated respectively in the cases T_1, T_2, T_3 we finally obtain
the circuit given.

The adder will be shown as a single block as in Fig. 11. The input with the
inhibiting circle being of course that shown at the top in the complete
diagram.

6.   Outline of Logical Control

A simple form of logical control would be a list of operations to be carried
out in the order in which they are given. Such a scheme can be made to
cover quite a number of jobs, e.g. calculations from explicit formulae, and
has been used in more than one machine. However it lacks flexibility. We
wish to be able to arrange that the sequence of orders can divide at various
points, continuing in different ways according to the outcome of the
calculations to date. We also wish to be able to arrange for the splitting up
of operations into subsidiary operations. This should be done in such a way
that once we have written down how an operation is to be done we can use it
as a subsidiary to any other operation.

[[15]]

These requirements can largely be met by having the instructions on a form of erasible memory, such as the delay lines. This gives the machine the possibility of constructing its own orders; i.e. there is always the possibility of taking a particular minor cycle out of storage and treating it as an order to be carried out. This can be very powerful. Besides this we need to be able ⟦15⟧ to take the instructions in an order different from their natural order if we are to have the flexibility we desire. This is sufficient.

It is convenient to divide the instructions into two types A and B. An instruction of type A requires the central arithmetic part CA to carry out certain operations. Such an instruction, translated from its symbolic form into English might run:

Instruction 491   A. Multiply the content of TS 23 by the content of TS 24 and store the result in TS 25. Then proceed to carry out the next instruction (i.e. No. 492).
Instructions of type [B] merely specify the number of the next instruction.

Instruction 492   B. Proceed with instruction 301.

We must now explain in more detail how it comes about that we can branch the sequence of instructions and arrange for subsidiary operations. Let us take branching first. Suppose we wish to arrange that at a certain point instruction 33 will be applied if a certain digit is 0 but instruction 50 if it is 1. Then we may copy down these two instructions and then do a little calculation involving these two instructions and the digit D in question. One form the calculation can take is to pretend that the instructions were really numbers and calculate

$$D \times \text{Instruction } 50 + (1 - D) \times \text{Instruction } 33.$$

The result may then be stored away, let us say in a box which is permanently ⟦16⟧ labelled 'Instruction 1'. We are then given an order of type B saying that instruction 1 is to be followed, and the result is that we carry out instruction 33 or 50 according to the value of D.

⟦17⟧   When we wish to start on a subsidiary operation we need only make a note of where we left off the major operation and then apply the first instruction of the subsidiary. When the subsidiary is over we look up the note and continue with the major operation. Each subsidiary operation can end with instructions for this recovery of the note. How is the burying and disinterring of the note to be done? There are of course many ways. One is to keep a list of these notes in one or more standard size delay lines (1024),

⟦16⟧

with the most recent last. The position of the most recent of these will be
kept in a fixed TS, and this reference will be modified every time a subsidiary is started or finished. The burying and disinterring processes are
fairly elaborate, but there is fortunately no need to repeat the instructions
involved each time, the burying being done through a standard instruction
table BURY, and the disinterring by the table UNBURY.   [[18]]

7. External Organs

(i) *General.* It might appear that it would be difficult to put information
into the calculator and to take it out, on account of the high speeds
associated with the calculator, and the slow speeds associated with mechanical devices; but this difficulty is not a real one. Let us consider for instance
the output organ. We will allow the mechanical part of the output organ to
work at whatever pace suits it, to take its own time in fact. However we will
require it to give out signals stating when it is ready to accept information.
This signal provides a gate for the feeding of the information out to the
output organ, and also signifies to the calculator that it may note that
information as recorded and proceed to feed out some more. The preparation for feeding the information out consists merely in transferring it from
dynamic storages onto trigger circuits.   [[19]]

In the case of the output arrangements we have the full power of the
calculator behind us, i.e. we can do the conversion of the information into
the required form as an ITO. In the case of the input organ we must go more
warily. If we are putting the instruction tables into delay lines, then when
the power has been turned off all memory will have been effaced, including
the instruction tables. We cannot use instruction tables to get the information back, because the instruction tables are not there. We are able to get
over this difficulty as will be seen below.

(ii) *Output organ.* The output will go on to 32 columns of some Hollerith
cards. All the 12 rows may be used. On the receipt of a signal from the
calculator a card will begin to pass through a punch or 'reproducer'.
Shortly before each row comes into position for punching a signal is sent
back to the calculator and trigger circuits controlling the punches are set
up. After the punching another signal is sent to the calculator and the
trigger circuits are cleared. The reproducer punch also gives a signal on the
final exit of the card. The circuit is shown in connection with CA (Fig. 26).

(iii) *Input organ.* Let us first describe the action of this without worrying
about the difficulty concerning absence of instruction tables. It is very

similar to the output organ in many ways. The input is from 32 columns and 12 rows of a Hollerith card. When the calculator is ready a card release signal goes out to the card reader and a card begins to pass through. As each row comes into position for reading a signal is sent back to the calculator, which then prepares to accept the output from the reader at the moment appropriate for sending it to its destination in the delay line. It is assumed that this destination is already decided by the calculator. A signal is sent back to the calculator on the final exit of the card.

Now let us consider what is done right at the beginning. Arrangements are made for setting into CI and CD a certain invariable initial order and IN. These state that the card is to be transferred into a particular delay line, and that the next order is to be taken from a particular spot, which will actually be in this same delay line. The information in this delay line can contain sufficient orders to 'get us started'. The first few orders obeyed will probably be to take in a few more cards. The information on these will later be sorted to its final destination. When the final instructions are in place it will be as well to 'read them back'.

[20]

Actually it has been arranged that the special initial order consists of 0 throughout so that there is no need to set it up.

(iv) *Binary-decimal conversion*. It is proposed to do binary-decimal and decimal-binary conversion as ITO.* This will be appreciably assisted by the fact that short multiplication is a CAO.[†]

(v) *Instruction-table cards*. It was explained in connection with the input organ that the instructions would be on cards, of whose columns all but 32 were available for external use. A proposed use of the 80 columns is suggested below, without proper explanation; the explanation comes later.

	Columns
Genuine input	41–72
Repeat of destination	26–40
Popular name of group	1–8
Detail figure (popular)	9–11
Instruction (popular)	12–25
Job number	73–77
Spare	78–80

* ITO = Instruction Table Operation.
[†] CAO = Central Arithmetic Operation.

Of these the genuine input has already been spoken of to some extent, and ⟦21⟧
will be spoken of again further. The job number and the spare columns do
not require explanation. The popular data describe the instruction in letters
and figures in a manner appropriate for the operator to appreciate quickly
if for instance the cards are listed. In this respect we might say that the
popular data is like a telephone number Mol 1380 whereas the genuine ⟦22⟧
input is like the pulses used in dialling: indeed we shall probably carry the
analogy further and really only distinguish 10 different letters, as is done on
automatic exchanges. The popular data have also another important function, which only appears when we consider that the same instructions will
be used on quite different jobs. If we were just to number the instructions
serially throughout all the instructions ever used on any job, then, in the set
of instructions actually used in any particular job there would be large gaps
in the numbering. Suppose now that these instructions were stored in the
DS with positions according to their numbers there would be a lot of
wasted space, and we should need elaborate arrangements for making use
of this space. Instead, when a new job appears we take the complete set of
cards involved and make a new copy of each of them; these we sort into the
order of popular group name and detail figure. We then renumber them
consecutively in the binary scale. This number goes into the columns
described as 'repeat of destination'. The renumbering may be done either
with a relay counter attached to a collater, or by interleaving a set of master
cards with the binary numbers in serial order. To complete the process we
have to fill in other instruction numbers in binary form into the genuine
input, e.g. if an instruction in popular form were "... and carry out
instruction Potpan 15" the genuine input will have to be of form "... and
carry out instruction 001101 ... 1" where 001101 ... 1 is the new number
given to Potpan 15 in this particular job. This is a straightforward sorting
and collating process.

It would be theoretically possible to do this rearrangement of orders
within the machine. It is thought however that this would be unwise in the
earlier stages of the use of the machine, as it would not be easy to identify
the orders in machine form and popular form. In effect it would be
necessary to take an output from the calculator of every order in both
forms.

8. Scope of the Machine

The class of problems capable of solution by the machine can be defined
fairly specifically. They are those problems which can be solved by human

clerical labour, working to fixed rules, and without understanding, provided that

(a) The amount of written material which need be kept at any one stage is limited to the equivalent of 5,000 real numbers (say), i.e. about what can conveniently be written on 50 sheets of paper.

(b) That the human operator, doing his arithmetic without mechanical aid, would not take more than a hundred thousand times the time available on the calculator, this figure representing the ratio of the speeds of calculation by the two methods.

(c) It should be possible to describe the instructions to the operator in ordinary language within the space of an ordinary novel. These instructions will not be quite the same as the instructions which are normally given to a computer, and which give him credit for intelligence. The instructions must cover every possible eventuality.

Let us now give real examples of problems that do and problems that do not satisfy these conditions.

Problem 1   Construction of range tables. The complete process of range-table construction could be carried out as a single job. This would involve calculation of trajectories by small arcs, for various different quadrant elevations and muzzle velocities. The results at this stage would be checked by differencing with respect to other parameters than time. The figures actually required would then be obtained by interpolation and these would finally be rearranged in the most convenient form. All of this could in theory be done as a single job. In practice we should probably be wiser to do it in several parts in order to throw less responsibility on to the checking arrangements. When we have acquired more practical experience with the machine we will be bolder.

It is estimated that the first job of this kind might take one or two months, most of which would be spent in designing instruction tables. A second job could be run off in a few days.

Problem 2   To find the potential distribution outside a charged conducting cube. This is a problem which could easily be tackled by the machine by a method of successive approximations; a relaxation process would probably be used. In relaxation processes the action to be taken at each major step depends essentially on the results of the steps that have gone before. This would normally be considered a serious hindrance to the mechanisation of a process, but the logical control of the proposed calculator has been designed largely with such cases in veiw, and will have no difficulty on this score. The problem proposed is one which is well within the scope of the machine, and could be run off in a few minutes, assuming it was done as one of a sequence of similar problems. It is quite outside the scope of hand methods.

Problem 3   The solution of simultaneous linear equations. In this problem   [[23]]
we are likely to be limited by the storage capacity of the machine. If the coefficients in the equations are essentially random we shall need to be able to store the whole matrix of coefficients and probably also at least one subsidiary matrix. If we have a storage capacity of 6400 numbers we cannot expect to be able to solve equations in more than about 50 unknowns. In practice, however, the majority of problems have very degenerate matrices and we do not need to store anything like as much. For instance problem (2) above can be transformed into one requiring the solution of linear simultaneous equations if we replace the continuum by a lattice. The coefficients in these equations are very systematic and mostly zero. In this problem we should be limited not by the storage required for the matrix of coefficients, but by that required for the solution or for the approximate solutions.

Problem 4   To calculate the radiation from the open end of a rectangular wave-guide. The complete polar diagram for the radiation could be calculated, together with the reflection coefficient for the end of the guide and interaction coefficients for the various modes; this would be done for any given wavelength and guide dimensions.

Problem 5   Given two matrices of degree less than 30 whose coefficients   [[24]]
are polynomials of degree less than 10, the machine could multiply the matrices together, giving a result which is another matrix also having polynomial coefficients. This has important applications in the design of optical instruments.

Problem 6   Given a complicated electrical circuit and the characteristics of its components, the response to given input signals could be calculated. A standard code for the description of the components could easily be devised for this purpose, and also a code for describing connections. There is no need for the characteristics to be linear.

Problem 7   It would not be possible to integrate the area under a curve, as the machine will have no appropriate input.

Problem 8   To count the number of butchers due to be demobilised in June 1946 from cards prepared from the army records. The machine would be quite capable of doing this, but it would not be a suitable job for it. The speed at which it could be done would be limited by the rate at which cards can be read, and the high speed and other valuable characteristics of the calculator would never be brought into play. Such a job can and should be done with standard Hollerith equipment.

[[25]]   *Problem 9*   A jig-saw puzzle is made up by cutting up a halma-board into pieces each consisting of a number of whole squares. The calculator could be made to find a solution of the jig-saw, and, if they were not too numerous, to list all solutions.

This particular problem is of no great importance, but it is typical of a very large class of non-numerical problems that can be treated by the calculator. Some of these have great military importance, and others are of immense interest to mathematicians.

[[26]]   *Problem 10*   Given a position in chess the machine could be made to list all the 'winning combinations' to a depth of about three moves on either side. This is not unlike the previous problem, but raises the question 'Can the machine play chess?' It could fairly easily be made to play a rather bad game. It would be bad because chess requires intelligence. We stated at the beginning of this section that the machine should be treated as entirely without intelligence. There are indications however that it is possible to make the machine display intelligence at the risk of its making occasional serious mistakes. By following up this aspect the machine could probably be made to play very good chess.

9. Checking

It will be almost our most serious problem to make sure that the calculator is doing what it should. We may perhaps distinguish between three kinds of error.

(1) Permanent faults that have developed in the wiring or components, e.g. condensers that have become open circuit.
(2) Temporary errors due to interference, noise reaching unexpected levels, unusual combinations of voltages at some point in the circuit, etc.

[[22]]

(3) Errors due to the use of incorrect instruction tables, or even due to mistaken views as to what the circuit should do.

It will be our intention to install monitoring circuits to detect errors of form (1) fairly soon. The ideal to aim at should be that each conceivable form of failure would give a different indication on the monitor. In practice we should probably simply localise the error to some part, e.g. an adder, which could be changed and then examined at leisure.

Errors of type (2) should not occur when the apparatus is in proper working order, however when a component is beginning to age its deficiencies will often show themselves first in this sort of way. For instance, if the emission of a valve in a Kipp relay circuit is beginning to fail it will eventually not pass on any of the pulses it should, but this will begin with some occasional failures to react. The worst of this can probably be eliminated by frequent test runs in which the conditions of H.T. volts, interference, etc., are all modified in a way calculated to accentuate the deficiencies of the components. Those which are rather down at heel may then be removed, and when the conditions are restored to normal there should be a good margin of safety. We cannot of course rely on this 100%. We need a second string. This will be provided by a variety of checks of the   [[27]] types normally employed in computing, i.e. wherever we can find a simple identity which should be satisfied by the results of our calculations we shall verify it. For instance, if we were multiplying polynomials algebraically we should check by taking a particular value for the variable. If we were calculating the values of an analytic function at equal intervals we should check by differencing. Most of these checks will have to be set up as part of the instruction tables, and the appropriate action to be taken will also be put into them. A few checks will be made part of the circuit. For instance, all multiplications and additions will be checked by repeating them module 255.

Incorrect instruction tables (3) will often be shown up by the checks   [[28]] which have been put into these same instruction tables. We may also apply a special check whenever we have made up a new instruction table, by comparing the results with the same job done by means of a different table, probably a more straightforward but slower one. This should eliminate all errors on the part of the mathematicians, but would leave the possibility of lost cards, etc., when the table is being used a second time. This may perhaps be corrected by running a test job as soon as the cards have been put into the machine.

                                                                    [[23]]

There are three chief functions to be performed by the checking. It must eliminate the possibility of error, help to diagnose faults, and inspire confidence. We have not yet spoken at all of this last requirement. It would clearly not be satisfactory if the checking system in fact prevented all errors, but nobody had any confidence in the results. The device would come to no better end than Cassandra. In order to inspire confidence the checking must have some visible manifestations. Certainly whenever a check fails to work out the matter must be reported by the machine. There would not be time for all checks which do work out to be reported, but there could be a facility by which this could be laid on temporarily at moments of shaken confidence. Another facility which should have a good effect on morale is that of the artificial error. By some means the behaviour of the machine is disturbed from outside, and one waits for some error to be reported. This could be managed quite easily. One could arrange to introduce an unwanted pulse at any point in the circuit. In fact of course we cannot do very much about checking until the machine is made. We cannot really tell what troubles of this kind are in store for us, although one can feel confident that none of them will be insurmountable. We can only prepare against the difficulties we can foresee and hope that they will represent a large percentage of the whole.

10.   Time-Table, Cost, Nature of Work, Etc.

The work to be done in connection with the machine consists of the following parts:

(1) Development and production of delay lines.
(2) Development and production of other forms of storage.
(3) Design of valve-elements.
(4) Final schematic circuit design of LC and CA.
(5) Production of the electronic part, i.e. LC and CA.
(6) Making up of instruction tables.
(7) External organs.
(8) Building, power supply cables, etc.

(1) Delay lines have been developed for R.D.F. purposes to a degree considerably beyond our requirements in many respects. Designs are available to us, and one such is well suited to mass production. An estimate of £20 per delay line would seem quite high enough.

(2) The present report has only considered the forms of storage which

are almost immediately available. It must be recognized however that other forms of storage are possible, and have important advantages over the delay line type. We should be wise to occupy time which falls free due to any kind of hold-up by researching into these possibilities. As soon as any really hopeful scheme emerges some more systematic arrangement must be made.

We must be ready to make a change over from one kind of storage to another, or to use two kinds at once. The possibility of developing a new and better type of storage is a very real one, but is too uncertain, especially as regards time, for us to wait for it; we must make a start with delay lines.

(3) Work on valve element design might occupy four months or more. In view of the fact that some more work needs to be done on schematic circuits such a delay will be tolerable, but it would be as well to start at the earliest possible moment.

(4) Although complete and workable circuits for LC and CA have been described in this report these represent only one of a considerable number of alternatives. It would be advisable to investigate some of these before making a final decision on the circuits. Too much time should not however be spent on this. We shall learn much more quickly how we want to modify the circuits by actually using the machine. Moreover if the electronic part is made of standard units our decisions will not be irrevocable. We should merely have to connect the units up differently if we wanted to try out a new type of LC and CA.

(5) In view of the comparatively small number of valves involved the actual production of LC and CA would not take long; six months would be a generous estimate.

(6) Instruction tables will have to be made up by mathematicians with computing experience and perhaps a certain puzzle-solving ability. There will probably be a great deal of work of this kind to be done, for every known process has got to be translated into instruction table form at some stage. This work will go on whilst the machine is being built, in order to avoid some of the delay between the delivery of the machine and the production of results. Delay there must be, due to the virtually inevitable snags, for up to a point it is better to let the snags be there than to spend such time in design that there are none (how many decades would this course take?). This process of constructing instruction tables should be very fascinating. There need be no real danger of it ever becoming a drudge, for any processes that are quite mechanical may be turned over to the machine itself.

The earlier stages of the making of instruction tables will have serious repercussions on the design of LC and CA. Work on instruction tables will therefore start almost immediately.

(7) Very little need be done about the external organs. They will be essentially standard Hollerith equipment with special mounting.

(8) It is difficult to make suggestions about buildings owing to the great likelihood of the whole scheme expanding greatly in scope. There have been many possibilities that could helpfully have been incorporated, but which have been omitted owing to the necessity of drawing a line somewhere. In a few years time however, when the machine has proved its worth, we shall certainly want to expand and include these other facilities, or more probably to include better ideas which will have been suggested in the working of the first model. This suggests that whatever size of building is decided on we should leave room for building-on to it. The immediate requirements are:

Room for 200 delay lines. These each require about 6 inches of wall space if they are to be individually accessible, and if this is partly provided by cubicle construction 300 square feet is probably a minimum. To this we might add another 100 square feet for the temperature correction arrangements.

Space for LC and CA. This is difficult to estimate, but 5 eight foot racks might be a reasonable guess and would require another 200 square feet or more. In the same room we would put the input and output organs which might occupy 40 square feet. We should also provide another 100 square feet for operators tables, etc. 400 square feet would not be unreasonable for this room.

Card storage room. We would probably keep a stock of about 100,000 cards, a very insignificant number by normal Hollerith standards. 200 square feet would be quite adequate.

Maintenance workshop We would do well to be liberal here. 400 square feet.

This total of 1400 square feet does not allow for the planning of operations, which would probably be done in an office building elsewhere, nor for the processing of Hollerith cards which will probably be done on machinery already available to us.

Cost It appears that the cost of the equipment will not be very great. An estimate of £20 per delay line would be liberal, so that 200 of these would cost us £4000. The valve equipment at £5 per inch of rack space might total £5000. The power supply might cost £200. The Hollerith equipment would be hired, which would be advantageous because of the danger of it going out of date. The capital cost of such Hollerith equipment even if bought would not exceed £2000. With this included the total is £11,200.

Part II   Technical Proposals

11.   Details of Logical Control   12.   Detailed Description of the Arithmetic Part (CA)   13.   Examples of Instruction Tables   14.   The Design of Delay Lines   15.   The Design of Valve Elements   16.   Alternative Forms of Storage

11.   Details of Logical Control

In this section we shall describe circuits for the logical control in terms of the circuit elements introduced in §5. It is assumed that §5, 6 are well understood.

The main components of LC are as follows:

(1)   A short storage (like a TS) called *current data* CD. This contains nothing but the appropriate *instruction number* IN, i.e. the position of the next instruction to be carried out.

(2)   A short storage called *current instructions* CI. This contains the instruction being or about to be carried out.

(3)   A tree for the selection of a particular delay line, with a view to finding a particular instruction.

(4)   Timing system for the selection of a particular minor cycle from a delay line.

(5)   Timing system for the selection of particular pulses from within a minor cycle.

(6)   Arrangements for controlling CA, i.e. for passing instructions on to CA.

(7)   Arrangements for the continual change of the contents of CD, CI.

(8)   Timing arrangements for LC itself.

(9)   Starting device.

Let us first describe the starting device. This merely emits pulses synchronously with the clock from a certain point onwards, on the closing of a

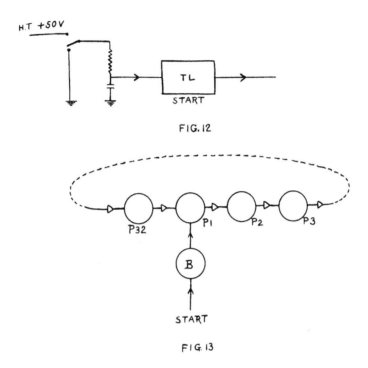

FIG. 12

FIG. 13

switch manually. The switch causes a voltage to rise and this eventually operates a trigger limiter. This starting mechanism sets a pulse running round a ring of valve elements providing the timing within a minor cycle. (Fig. 12, 13).

In order to check that this circuit is behaving we compare P32 with a signal which should coincide with it and which is obtained in another way, stimulating an SOS signal when there is failure. This forms one of the monitoring devices. We are not showing many of them in the present circuits. (Fig. 14).

The timing system for the selection of minor cycles is quite simple, consisting chiefly of a 'slow counter' SCA, which counts up to 255 in the scale of 2, keeping the total in a delay line of length 8. The pulses counted are restricted to appearing at intervals which are multiples of eight. As shown (Fig. 15) it is counting the pulses P10. The suppression of the outputs at P9 prevents undesirable carries from the most significant digit to the least.

The information in CD and CI being in dynamic (time) form is not very

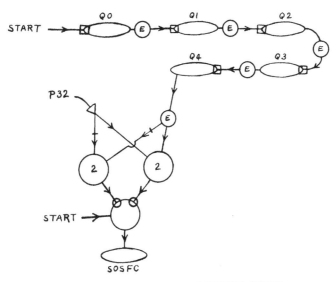

FIG. 14   A CHECKING CIRCUIT

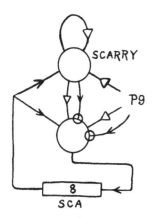

FIG. 15   SLOW COUNTER SCA

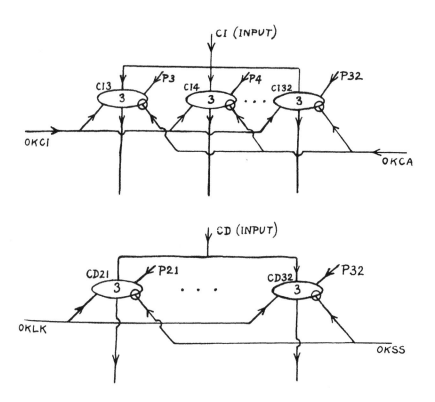

FIG. 16 STATICISERS FOR CI, CD.

convenient for control purposes. We therefore convert this information into static form, i.e. we transfer it on to trigger circuits. (Fig. 16).

It will be convenient to make use of a symbolic notation in connection with the valve circuits. We write A & B (or manuscript $A ⅋ B$) to mean 'A and B'. If A and B are thought of as numbers 0 or 1 then A & B is just AB. We write $A \vee B$ for 'A or B'. With numbers $A \vee B$ is $1 - (1 - A)(1 - B)$. We also write $-A$ (manuscript $\sim A$) for 'not A' or $1 - A$. Other logical symbols will not be used. Where a whole sequence of pulses is involved, it is to be understood that these operations are to be carried out separately pulse by pulse. We shall combine these symbols with the symbol + which refers to the operations of the adder. Thus for example $(A + (P3 \vee P4))$ & $-P5$ means that we take the signal A and add to it a signal consisting of pulses in positions 3 and 4 and nowhere else, (addition in the sense of the

[30]

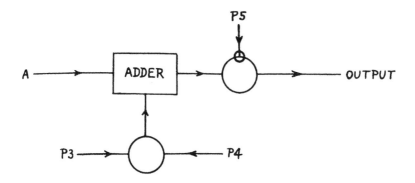

FIG. 17 ILLUSTRATING A CONVENTION

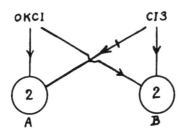

FIG. 18

adder circuit), and that we then suppress any pulses in position 5, as in Fig. 17. We will also abbreviate such expressions as P5 ∨ P6 ∨ P7 ∨ ⋯ P19 to P5–19, and expressions such as A & P14–18 to A 14–18.   [[29]]

In circuit diagrams we have the alternatives of showing the logical combinations by formulae or by circuits. There is little to choose but there may be something to be said for an arrangement by which purely logical combination is not shown in circuit form, in order that the circuits may bring out more clearly the time effects.

We have agreed that there shall be two kinds of instructions, A and B. These are distinguished by CI 3. The standard forms for the two types of instructions are:

Type A   Carry out the CA operations given by digits CI 5–32, and construct a new CD according to the equation CD = (CD′ + P19) & −P17.

[[31]]

Type B   Construct a new CD according to the equation CD = CI 17–32. Pass the old CD into TS 13.

CD′ here represents the old CD. The significance of the formula for CD in case A is this. Normally it is intended that after an operation of type A the next instruction to be followed will be that with the next number, and it might be supposed therefore that the formula CD = CD′ + P17 would apply. Actually we deviate from this simple arrangement in two ways. Firstly we find it convenient to have a facility by which an instruction may be taken from a TS, viz. TS 6: this has considerable time saving effects. The convention is that a digit 1 in column 17 indicates that the next instruction

[[30]]   is to be taken from TS 6. This will involve our having only the digits CI 18–32 available to indicate normal positions for instructions and would suggest that the formula should be CD = CD′ + P18. However if we did this we should always be obliged to have orders of type B in TS 6, for if we

[[31]]   had an order of type A we should find that we had to go on repeating that order. If however we have the formula CD = (CD′ + P18) & −P17 we can obey an instruction in TS 6 and then revert to the instruction given by CI 18–32; a much more convenient arrangement. It remains to explain why we have P19 rather than P18. This is due to the fact that we wish to avoid the necessity of waiting a long time for our instructions. If the equation were the one with P18 it would mean that the next instruction to be obeyed, after one of type A, is always adjacent to it in time. This would mean that even with the shortest CA operations the next instruction would have gone by before we were ready to apply it; we should always just miss the boat. By putting P19 instead of P18 we give ourselves an extra minor cycle of time which is normally just what we need. In order that the consecutive instructions may be consecutively numbered in spite of this it is best to adopt

[[32]]   a slightly unconventional numbering system for the minor cycles (see Fig. 19).

A number of trigger circuits are employed to keep track of the stages which the various processes have reached at any moment. The most important of these are listed below with a short description of the functions of each.

OKCI   This is stimulated when the new instruction has been found and is available at the input of CI, and the CA operations belonging to the last instruction have been carried out. Stimulation begins simultaneously with stimulation of P1, and ends on a P32. The end of OKCI has to wait for the gating of CD, indicating that the new CD is available at its input.

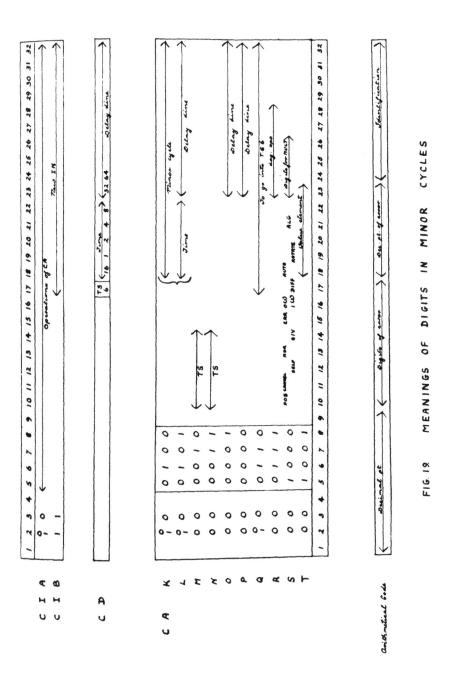

FIG 19.   MEANINGS OF DIGITS IN MINOR CYCLES

OKCA Only applies in case A and indicates that the CA operations have been finished.

OKCK Indicates that we may now begin to look for the next instruction with a view to putting it into CI. It is stimulated when OKCI is extinguished, and is itself extinguished when the new CI has been found.

[[33]] We may now describe the time cycle of LC. Let us begin at the point where OKLK is stimulated indicating that the search for the new CI may now begin, because we have finished with the old one and information for finding the new one is now available in CD. The new CI is determined by digits 17–32 of CD. Of these digits 23–32 determine the delay line and 18–22 determine the minor cycle within the delay line. A digit 1 in column 17 indicates that the order is to be taken from TS 6 instead of from the longer delay lines. This digit is erased whenever we obey an instruction of type A. Digits 23–32 are set up on trigger circuits and operate via trees as described below. Digits 18–22 determine the time at which we must take the output of the delay line. We compare these digits with the output of the slow counter SCA (Fig. 15) and when they agree we know that the right moment has come. It is convenient to arrange that the slow counter is always one minor cycle ahead of time, so as to give us time to organise ourselves before taking the required output. As has been mentioned the order of the digits in CD is arranged rather unconventionally in order to put consecutively numbered minor cycles in alternate positions; this has time saving effects. The required minor cycle now passes into CI and the signal OKSS is given; OKLK is suppressed. When the CA operations belonging to the last instruction have been finished OKCA is stimulated and with it OKCI. We are now able to initiate any new CA operations (case A) and to set up the new CD. When this has been done we have finished with CI and suppress OKCI, which automatically stimulates OKLK beginning the cycle over again. (Figs. 22, 22a).

The digits 23–32 determine the delay line required. This amounts to 10 digits and will certainly be adequate for our present programme. Treeing is done in two stages, going first through trees for three or four digits only. These are TRA 000...TRA 111, TRB 000...TRB 111, TRC 0000, ..., TRC 1111. These number 32 valve elements. At the second stage there are 1024 valve elements TREECI 0000000000, ... TREECI 1111111111. The connections are shown for TREECI 1011101101. The connection from CI 17 prevents any of the TREECI elements being stimulated when CI 17 is

[[34]]

stimulated. This is required to deal with the case where the next order is taken from TS 6 and not from the delay lines. (Fig. 20).

It is very probable that some other form of tree circuit, not capable of being drawn in terms of our valve elements, will be used, and the same will apply to many parts of the circuit. It is thought worth while however to draw these circuits, if only to clarify what it is intended the circuits should do.

We have a similar tree system for the selection of temporary storages.

12. Detailed Description of the Arithmetic Part (CA)

We shall divide the CA operations into a number of types. We shall make provision for 16 types, but for the present will only use nine. The types are distinguished by digits CI 5–8.

Type K   Pass the content of TS 6 into a given minor cycle.

Type L   Pass the content of a given minor cycle into TS 6.

Type M   Pass the content of a given TS into TS 6.

Type N   Pass the content of TS 6 into a given TS other than TS 4 or TS 5,   [[34]] or TS 8 or TS 1.

Type O   Pass the content of the first 12 minor cycles of a given DL out   [[35]] onto a card via the reproducer.

Type P   Pass the content of the card at present in the card reader on to a given DL.

Type Q   Pass CI 17–32 into TS 6.

Type R   Various logical operations and others yielding results forming one minor cycle, to be performed on the contents of TS 9 and TS 10 and transferred to TS 8.

Type S   Arithmetical operations yielding a result requiring more than one minor cycle for its retention. Results go into TS 4 and TS 5.

Type T   Stimulate a given valve element.

A trigger circuit is associated with each type. With the exception of Q these are all excited for a period consisting of a number of complete minor cycles beginning with a P1 and ending with a P32.

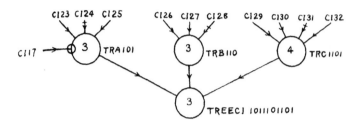

FIG. 20   A TREE

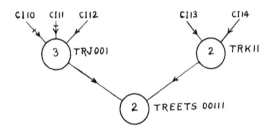

FIG. 21

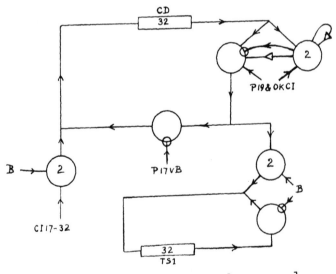

FIG. 22   CIRCUIT FOR CD [PART OF LC]

[36]

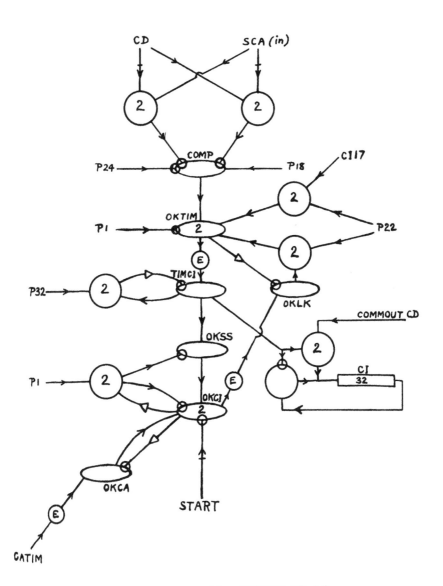

FIG. 22 a. TIMING OF LC

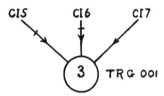

FIG. 23

The main components of CA are the 32 temporary storages TS 1–32. Of these TS 1–12 have some special duties.

⟦36⟧   TS 1 is used to carry the retiring data, i.e. the CD which applied just before the last instruction of type B.

TS 2 and TS 3 contain the arguments for the purely arithmetical operations, or most of them, and for the logical operations.

TS 4 and 5 contain the results of the arithmetical operations. They are frequently connected up in series to form a DL 64. This is because the results of most of the arithmetical operations are sequences of more than 32 but not more than 64 digits.

TS 6 is used as a shunting station for the transfer of information from place to place.

TS 7 is used to carry the digits of a number m when it is proposed to multiply by 2^m.

TS 8 is used to carry the result of logical operations and other operations not requiring more than one minor cycle.

TS 9 and TS 10 are the inputs for the logical operations.

TS 11 will usually be used in connection with error calculations, and accordingly has a special role in the production of multipliers.

TS 12 is used for the timing in 'automatic' multiplication and for the selection of unusual combinations of digits in the multiplier. The word 'automatic' is used because of an analogy from desk machines.

To decide between types K to T we use CI 5–8. Digits 5, 6, 7 are treed out to the valve elements TRG 000, ... TRG 111, as in Fig. 23. These tree elements are each associated with two types, which are distinguished by CI 8. Thus TRG 000 would be identical with K ∨ L, if it were not for timing. For this timing we introduce CATIM which is to be stimulated during the appropriate time in CA operations. K ∨ L is identical with TRG 000 & CATIM (Fig. 24).

⟦38⟧

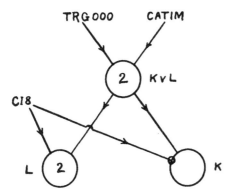

FIG. 24

In case K we pass the output of TS 6 to COMMIN and hence to the inputs of all the delay lines. We gate the appropriate one of these at the appropriate time, given by TIMCA by comparison of the output of the slow counter SCA with CI.

In case L we do somewhat similarly, passing the appropriate output to COMMOUT and thence to the input of TS 6 at the appropriate time given by TIMCA.

In case M we gate the appropriate output and pass into TS 6.

In case N we pass the output of TS 6 to the inputs of the other TS, only gating the one required.

In case O the first effect is to set the mechanism in motion to pass a card through the reproducer. By means of a commutator arrangement or otherwise the reproducer sends back a series of pulses which indicate the times when the reproducer punches are ready to accept current. In the circuit diagram (Fig. 25) two sets of pulses are shown which are intended to mark the beginnings and ends of these periods. They may be separately provided by the reproducer, or one may be derived from the other by delaying or otherwise. The two sets of pulses each control trigger limiters connected up so as to extinguish one another. (Do not confuse this with the two mutually extinguishing triodes that will normally form part of a trigger circuit or trigger limiter). One of the trigger limiters TIMOUTCARD stimulates the trigger circuit OUTIM on the first admissible P 10. A pulse on the stimulation of OUTIM goes into a slow counter SCB and enables us

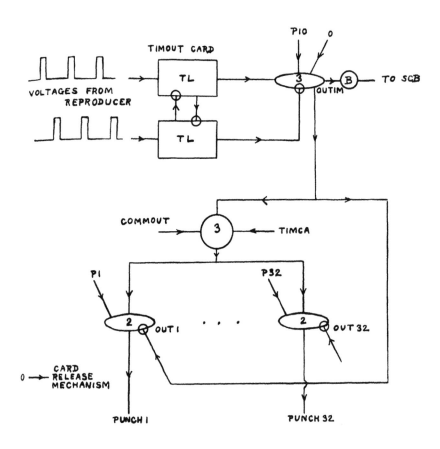

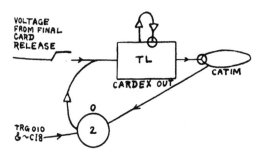

FIG. 25 OUTPUT CIRCUIT

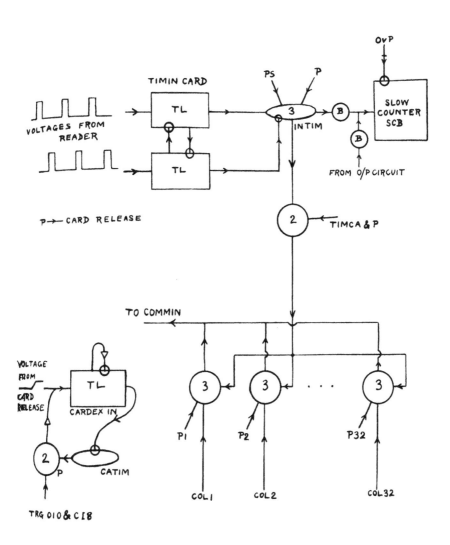

FIG. 26 INPUT CIRCUIT

to keep track of the number of rows of the card that have been punched. The content of SCB is compared with that of SCA and when they agree we know that the minor cycle which we wish to pass out is now available, and TIMCA is accordingly stimulated. TIMCA and OUTIM together permit COMMOUT to pass out to the trigger circuits OUT 1 . . . OUT 32 on which it is set up statically and controls the punches.

On the final exit of the card the reproducer sends back a signal to the calculator, which, in combination with O operates a trigger limiter CARDEXOUT. This suppresses CATIM and hence O. CARDEXOUT has feedback to suppress itself, and this will be successful because O will have been suppressed by the time it comes to act.

The behaviour in case P (input) is very similar. The chief difference is that whereas OUTIM was used to gate the output from the calculator INTIM is used to gate the input.

It should be noticed that a completely blank instruction has a definite meaning, viz. to pass the material on the card in the reader into DL 0000000000.

In Fig. 27 TS 01101 typifies any of the TS as regards output connections shown on other diagrams. It is also typical as regards input connections, except as regards TS 4, 5, 8, 1, which have no input connections except those shown on other diagrams.

In the case of operations of type R we shall calculate all of the expressions involved and select them by means of tree elements, digits 18 to 23 being used. The operations so far are:

Digits 000000   TS 8 = TS 9 & TS 10.
Digits 001000   TS 8 = TS 9 $\vee$ TS10.
Digits 010000   TS 8 = $-$ TS 10.
Digits 011000   TS 8 = (TS 9 & TS 10) $\vee$ ($-$ TS 9 & $-$ TS 10).
Digits 100000   TS 8 = 0.

As we shall have very much to say about type S we shall make a few remarks first about type T. In order to be able to obtain a rather direct access from the instructions to the valves we shall introduce a number of valve elements which can be stimulated to order. We may have 64 of these, say FLEX 000000 to FLEX 111111. The circuit will be simply as shown in Fig. 31. It is intended that the outputs of these valve elements should be connected in various ways into the circuit when it is desired to try out new circuit arrangements. It is thought that they may often provide means for

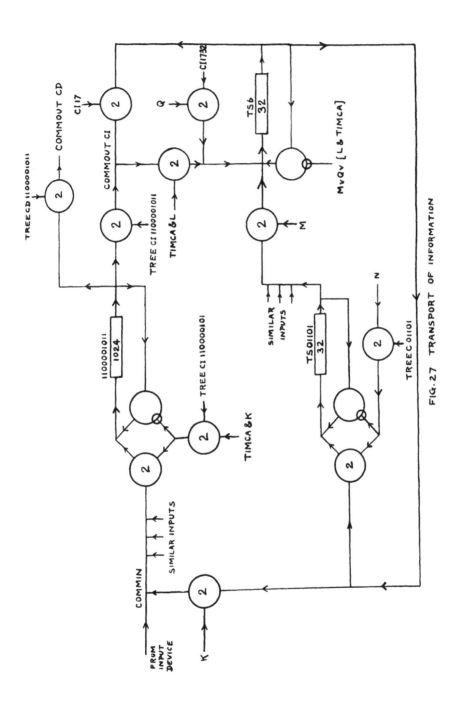

FIG. 27 TRANSPORT OF INFORMATION

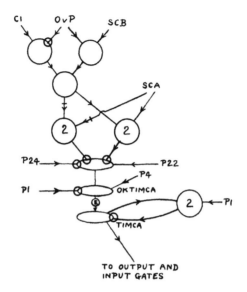

FIG. 28 MINOR CYCLE SELECTION FOR CA.

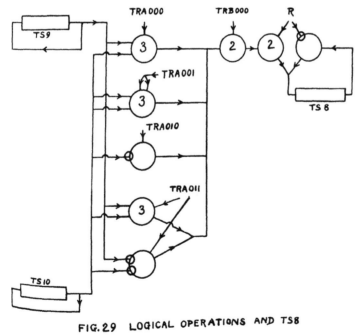

FIG. 29 LOGICAL OPERATIONS AND TS8

[[44]]

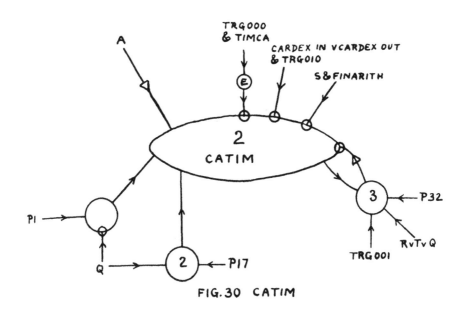

FIG. 30   CATIM

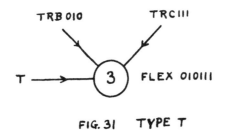

FIG. 31   TYPE T

doing things simply which could be done lengthily as an ITO. To an extent this represents a compromise between the new system of 'control by paper' and the old plugboard and soldering-iron techniques.

We shall also describe the timing arrangements before passing on to type S. We have already mentioned CATIM which determines the timing but we have still to mention what controls CATIM. CATIM is stimulated as soon as the first P1 appears after the signal A, or, in case Q, the first P17. It is extinguished by a variety of means. In cases K and L it is extinguished by the ending of TIMCA indicating that the required minor cycle has just passed through. In cases M, N, R, T, it is only permitted to last for one minor cycle. In case Q it is also only allowed to last for half a minor cycle. In cases O, P the extinguishing signal is CARDEX, which is given by the card reproducer of reader on the final exit of the card, via a trigger-limiter. In case S the signal comes from FINARITH.

The facilities provided under type S are not easily enumerated, because they do not consist of a number of different operations stimulated by different tree valve elements, as for instance applies in the case of the logical processes. Rather they are to be thought of as one process which can be modified in various ways. The standard process always involves converting the content of TS 4 and TS 5 into 'series form', i.e. instead of connecting the outputs of TS 4 and TS 5 to their own inputs they are connected to each others. When they are so connected their content will be described as the 'partial sum'. Some quantities are then added to or subtracted from the partial sum. If the quantity is to be added then POS is stimulated, otherwise they are subtracted. We may if we wish cancel the original partial sum before adding in which case we must stimulate CANCEL for a period of two minor cycles. The quantity to be added or subtracted is expressible as the product of a quantity known as the 'multiplicand' and an integer which may be taken to lie in the range -7 to 15, positive values being the more normal. The multiplicand may be taken from TS 3 or from the partial sums register itself. This latter case is convenient for the purpose of multiplying the partial sum by a small integer without a complicated series of previous transfers; if the multiplicand is taken from the partial sums register then SELF is stimulated. The multiplier may also be taken from a variety of sources. It may be taken from TS 2 or from CI or from TS 11, and we accordingly stimulatie NOR, GIV or ERR. The multiplier consists of four consecutive digits from whichever source is chosen. The choice of the digits is made by means of a choice of one of the pulses P1 to P32 to enter on a certain line (DIGIT). At present it is suggested that in case NOR this should be P1, resulting in the use of digits 1, 2, 3, 4, in case IV   it should be P23 resulting in the use of digits 23, 24, 25, 26, in case ERR1 it should be P10, and in case ERR2 it should be P14. In case DIFF these arrangements are to be overridden and the pulse will be stored in TS 12 and taken from there.

In case AUTO the above fundamental process is repeated eight times. In each repetition the multiplicand is taken from TS 3, but it is modified each time by multiplication by 2^4, this effect being obtained by allowing it to circulate in a DL34 during AUTO. We also wish to take different digits of the multiplier at each repetition of the process; this is done by taking our pulse from TS 12 but allowing it to circulate in a DL 34 also. Facilities are also provided for multiplying the partial sum by a power of 2. Although the circuits are arranged so that this could be combined with other operations, it is not intended that this should be done. The facility consists in enabling the partial sums to be delayed by any time up to 63 and passed through for a period of 2 or 3 minor cycles as desired. The amount of delay is taken from digits 1–5 of TS 7.   We stimulate ROTATE 2 or ROTATE 3 according as we wish the rotation to last for 2 or 3 minor cycles.

It may be as well to describe how some rather definite operations are done.

Addition   We do not have a facility for addition of two given numbers so much as for the addition of a given number into the partial sum. To add the content of TS 3 into the partial sum we must stimulate S, POS, GIV, and must also set up the number 1 in columns 24–27. The multiplicand is then TS 3 and the multiplier is 1.

Subtraction   As addition but we do not stimulate POS.

Short multiplication (A)   To multiply TS 3 by 6 (say) proceed as for addition with 0110 in columns 24–27 instead of 1000. We shall very likely also want to cancel the original content of the partial sums register and therefore stimulate CANCEL.

Short multiplication (B)   To multiply the partial sum by 6 we must stimulate S, POS, CANCEL, SELF, GIV, and set up 0110 in CI 24–27.

Short multiplication (C)   As B but do not cancel and put 1010 in CI 24–27.

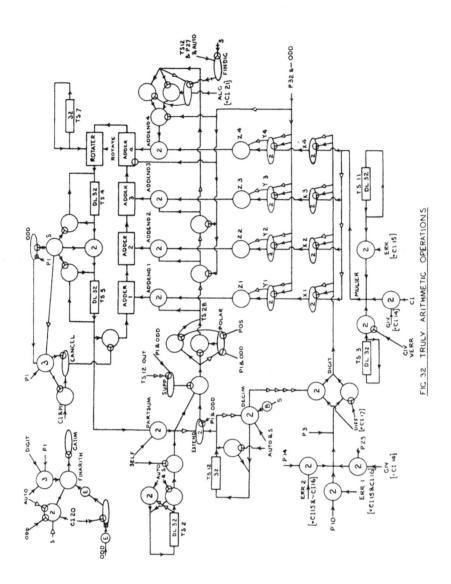

FIG 32 TRULY ARITHMETIC OPERATIONS

Short multiplication with addition   We wish to multiply TS 3 by TS 2 and add into the partial sum. We stimulate POS, NOR, AUTO, DIFF.

Long multiplication with subtraction   If we wish to subtract from the partial sum we do not stimulate POS.

Division is an ITO and will probably be carried out by means of the recurrence relation $u_0 = 3/4$, $u_{n+1} = u_n(2 - au_n)$. The limit of the sequence u_n is a^{-1} provided $1 < a < 2$.

The appropriate instructions for these operations will be found in Fig. 37.

The content of TS 2 or TS 3 is best considered to be a binary integer, i.e. that the least significant digit is in the units position. We must also consider that the most significant digit has reversed sign. The least significant digit appears at time P1 and the most significant at P32. In the partial sums register similarly the least significant digit is to be considered to be in the units position and the most significant to have reversed sign and to appear 63 pulses later. In order to keep track of which part of the partial sum is available at any moment we have a signal ODD which is stimulated during the first minor cycle of the stimulation of S, and thereafter in alternate minor cycles so long as S is stimulated. When the multiplicand is taken from TS 3 we have to make some slight modifications to it before it is in suitable condition for adding into the partial sum. We have to convert the periodic signal with period 32 or 34 into a sequence of 64 digits of which 32 form the original content of TS 3, and the rest is a sort of padding. We may call the 32 digits the *genuine digits*. Those digits of padding which are less significant than the genuine digits are to be all zero, those which are more significant are to be the same as the most significant genuine digit. It will be seen that this modified multiplicand MUCAND 2 has the same meaning as the original multiplier, but expressed in the code which is appropriate to the partial sum, and multiplied by the power of 2 which is required at the time. It may be necessary to change the sign of this multiplicand, if POS was not stimulated. A simple circuit will do this (Fig. 34).

Owing to the fact that the partial sums register is a closed cycle of 64 there is a danger of carries from the most significant digit on to the least significant. This has to be prevented, and it is done by suppressing the carry in the appropriate adder at the time P32 & −ODD. This is shown by an inhibiting connection on to the adder.

The detailed correctness of the circuits is best verified by working

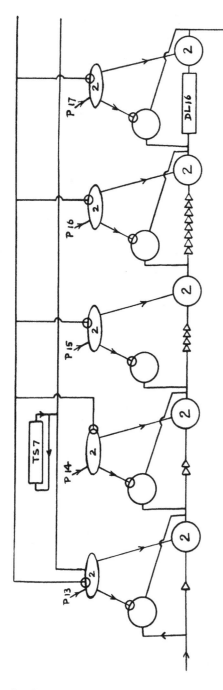

FIG. 33 THE ROTATER

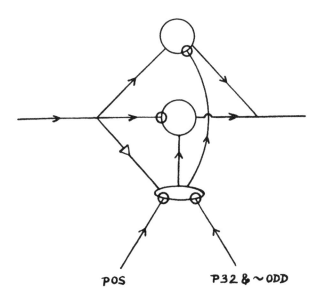

POS   P32 & ~ODD

FIG. 34. SIGN CHANGER

through various particular cases. It is necessary to work several different
ones in order to bring out the various different special points involved. In
Fig. 35 the preliminaries to a long multiplication have been worked. This
shows the setting up of the new CI and the transfer of digits to the valve
elements Z1, Z2, Z3, Z4. It brings out the point of adding 2 rather than 1 to
the CD in cases A, B, for we are just in time to catch the next instruction.
The final stages of the multiplication are shown in Fig. 36. Here it has been
assumed that the minor cycle is of length 16, in order to reduce the space
occupied by the working.

13.   Examples of Instruction Tables

In this chapter a short account of the paper technique of using the machine
will be given. I shall try to give some idea of what the instruction tables for a
job will be like and how they are related to the job and to the machine. This
account must necessarily be very incomplete and crude because the whole
project as yet exists only in imagination.

Each instruction will appear in a number of different forms, probably
three or four.

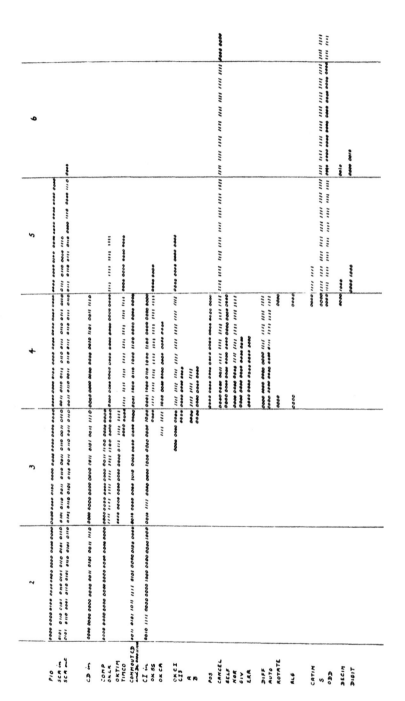

FIG. 35.   LC PRELIMINARIES TO A MULTIPLICATION

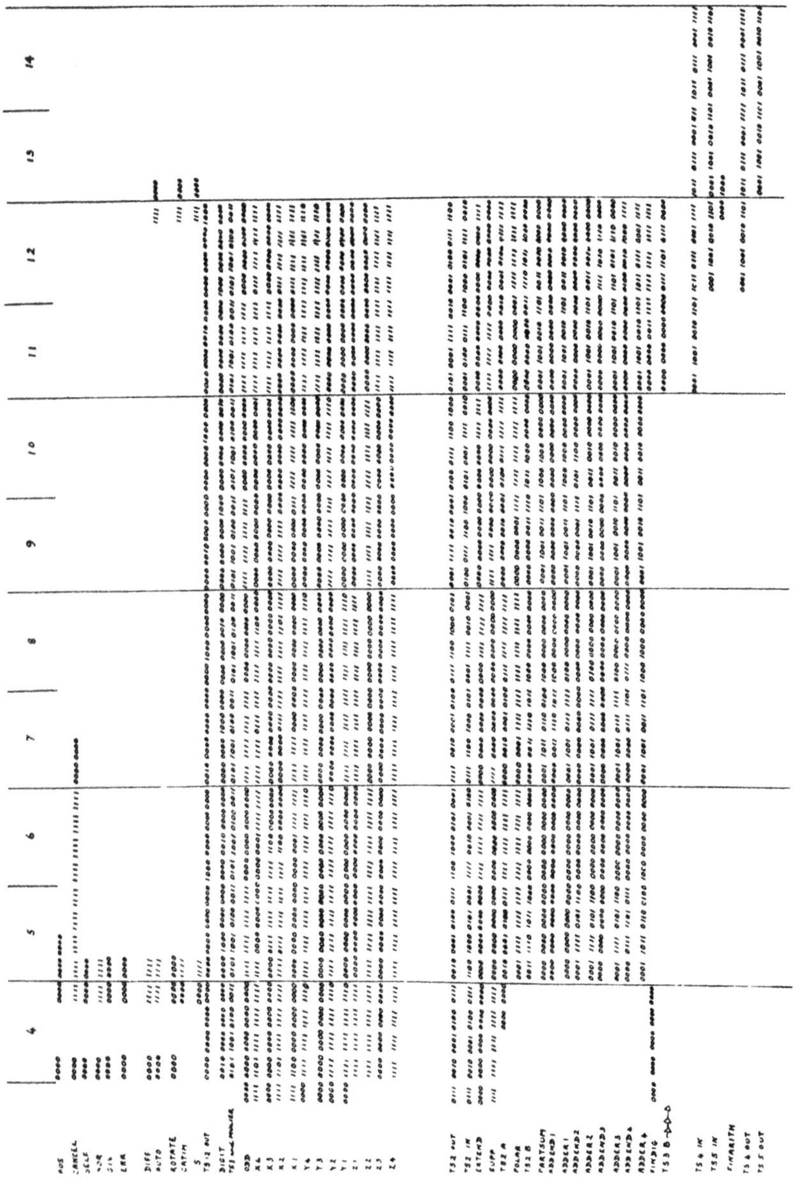

FIG. 36.   A MULTIPLICATION

	1	2	3	4	5	6	7	8	9	10 POS CANCEL	11 SELF	12	13 NOA	14 GIV	15 ERA	16 1	17 2 DIFF	18 AUTO	19 ROTATE	20	21	22	23	24	25	26	27	28	29	30	31	32
addition			0	1	1	1	0	0		1	0	0	0	1	0	1	0	0	0					1	0	0	0					
subtraction			0	1	1	1	0	0		0	0	0	0	1	0	1	0	0	0					1	0	0	0					
Short Mult. A (×6)			0	1	1	1	0	0		1	0	1	0	1	0	1	0	0	0					0	1	1	0					
Short Mult. B (×6)			0	1	1	1	0	0		1	1	1	0	1	0	1	0	0	0					0	1	1	0					
Short Mult. C (×6)			0	1	1	1	0	0		1	0	1	0	1	0	1	0	0	0					1	0	1	0					
long Mult. + add			0	1	1	1	0	0		1	0	0	1	0	0	0	0	1	0													
long Mult. + subtract			0	1	1	1	0	0		0	0	0	1	0	0	0	0	1	0													

FIG. 37.   INSTRUCTIONS FOR CERTAIN ARITHMETICAL OPERATIONS

[[54]]

Machine form   When the instruction is expressed in full so as to be understood by the machine it will occupy one minor cycle. This we call machine form.

Permanent form   The same instruction will appear in different machine forms in different jobs, on account of the renumbering technique as described on p. 38. Each of these machine form instructions arises from the permanent form of the instruction. These permanent forms are on Hollerith cards and are kept in a sort of library.

Popular form   Besides the cards we need some form of the table which can be easily read, i.e. is in the form of print on paper rather than punching. This will be the popular form of the table. It will be much more abbreviated than the machine form or the permanent form, at any rate as regards the descriptions of the CAO. The names of the instructions used will probably be the same as those in the permanent form.

In addition to these we must recognise the 'general description' of a table. This will contain a full description of the process carried out by the machine acting under orders from this table. It will tell us where the quantities or expressions to be operated on are to be stored before the operation begins, where the results are to be found when it is over and what is the relation between them. It will also tell us other important information of a rather dryer kind, such as the storages that must be left vacant before the operation begins, those that will get cleared or otherwise altered in the process, what checks will be made, and how various possible different outcomes of the process are to be distinguished. It is intended that when we are trying to understand a table all the information that is needed about the subsidiaries to it should be obtainable from their general descriptions.

The majority of actual instruction tables will consist almost entirely of the initiation of subsidiary operations and transfers of material. It should be recognised however that the time spent will be in quite different proportions. The three most time consuming operations are multiplication, waiting for material in long delay lines, and transfers of material. In some jobs the input and output of material may also be very time-consuming.   ⟦37⟧

In order to give a fairly complete picture of what the tables are like I am giving examples of two tables, of which one is elementary and does not involve subsidiaries; the other is a more advanced table and consists largely   ⟦38⟧
of such orders. Besides these I have added a number of general descriptions of tables.

⟦55⟧

The fundamental table chosen is INDEXIN, used for finding a minor cycle whose position has been written down in a particular place.

In these tables DL *m*, *n* will denote the *n*th minor cycle of DL *m*.

INDEXIN (general description)   The minor cycle whose position is described in digits 17–32 of TS 27 is transferred to TS 28. The contents of TS 2, 3, 4, 5, 6, 8, 9, 10 get altered in the process.
Now follows the popular form of the table.

INDEXIN

1	Q, 0000, 0100, 0000, 0000	2
2	TS 6–TS 2	3
3	ADD 'A'	4
4	ROTATE 16	5
5	TS 4–TS 6	6
6	TS 6–TS 9	7
7	TS 27–TS 6	8
8	TS 6–TS 10	9
9	OR	10
10	TS 8–TS 6	11
11	B, 1, INDEXIN 11	
12	TS 6–TS 28	13
13	B, BURY	

[[39]]
[[40]]   The first column gives the popular form of the name of the instruction, and the last column that of the next instruction to be followed. In most cases this could in theory be omitted because of the instructions being of type A. When the instructions are of type A the middle column describes them in abbreviated form. For instance TS 6–TS 3 describes the operation
[[41]]   of transferring the content of TS 6 into TS 3. Expressions of form Q, ... mean an instruction of type Q, and the expression after the comma describes what is in columns 17–32. ADD 'A' is to mean 'Add TS 2 into TS 4 cancelling the partial sums', ROTATE 16 means 'Rotate the content of TS 4, TS 5 forwards 16 places', OR is a logical operation.

The expression B, 1, INDEXIN 11 is intended to stand for B in column 3, 1 in column 17 and INDEXIN 11 in columns 17–32.

Outline of operation (INDEXIN)   From 1 to 10 we are constructing the instruction which tells us to make the appropriate transfer and putting that instruction into TS 6. The instruction B, 1, INDEXIN 11 requires us to

carry out the instruction in TS 6. The new IN formed will be 0, INDEXIN   [42]
12 so that we then continue with instruction INDEXIN 12.

The table for INDEXIN is shown in full in Fig. 38.   [43]

We use the convention that no digit is shown if the value of the digit is not
significant. Both 0 and 1 are shown if either value is possible, and
significant.

DISCRIM (general description)   If TS 8 contains any digit 1 then TS 15 is   [44]
passed into TS 24, otherwise TS 16 is passed into TS 24. The contents of TS
2, TS 3, TS 4, TS 5, TS 8 are altered.
Outline of operation   TS 8 is transferred to TS 2 and then subtracted from
zero, passing into the partial sums register TS 4, TS 5. By taking out TS 5 we
obtain a minor cycle full of digits 1 or of digits 0 according as there was or
was not a digit 1 in TS 8 originally. We then form (TS 5 & TS 15) $\vee$ ($\sim$ TS 5
& TS 16) by logical operations and pass it on to TS 24.

This table provides the main means of deciding between two alternative
procedures, by setting up one or the other of two instructions, contained in
TS 15 or TS 16.

PLUSIND (general description)   1 is added to the position reference in TS
27, e.g. DL 7, 9 becomes DL 7, 10, but DL 7, 32 becomes DL 8, 1.

TRANS 45 (general description)   The following set of transfers is made

TS 22–TS 20,   TS 23–TS 21.

BURY (general description)   The content of TS 1 with 1 added is trans-   [45]
ferred to the position indicated in TS 31, and 1 is added to the reference in
TS 31. We then proceed to carry out the instruction in TS 1.

UNBURY (general description)   The minor cycle whose position is given   [46]
in TS 31 is taken to be position of the next instruction.

MULTIP (general description)   The number in TS 18, 19 is multiplied by
the number in TS 20, 21: the result is brought to standard form by shift of
decimal point. An error is obtained for the product by using the errors in
the given numbers and allowing for rounding off. The result is stored in
TS 22, 23.

ADD is analogous to *MULTIP*.

As an example of a more complicated process, I have chosen the calculation of the value of a polynomial.

[57]

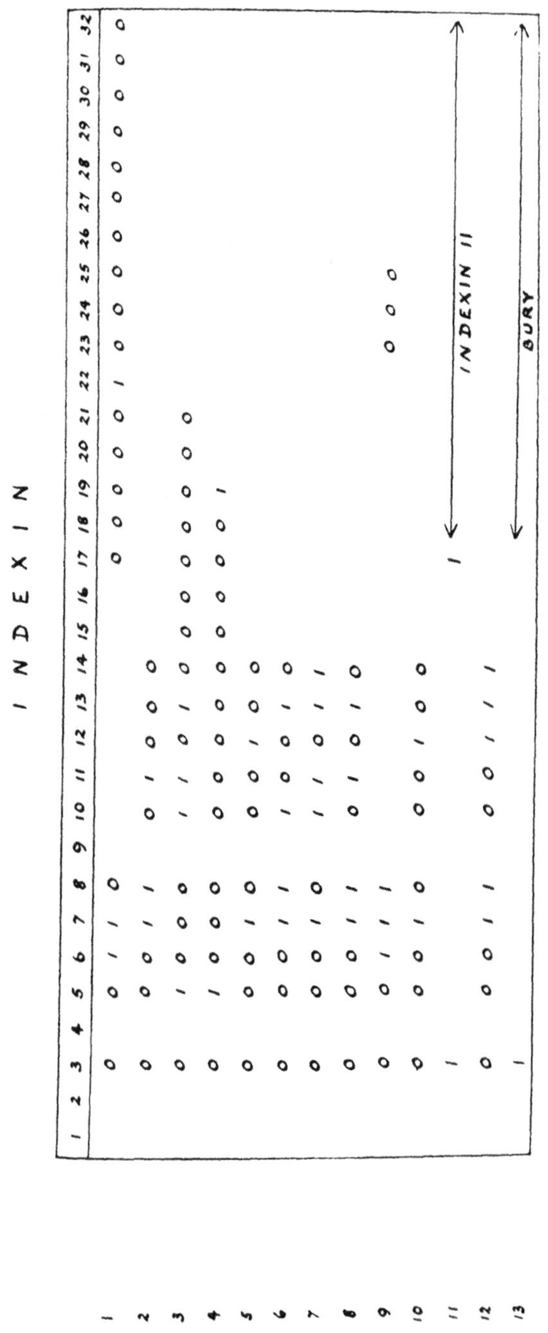

FIG. 38.   INSTRUCTION CARDS FOR INDEXIN (genuine Input)

CALPOL (general description)   The minor cycles of DL 3 taken in pairs conatin the coefficients of a polynomial in descending order. Evidently we are restricted to degrees not exceeding 15, and we assume the degree always to be 15, filling up with appropriate zero coefficients. The value of this polynomial will be calculated for the argument in TS 13, TS 14 and the result will be transferred to TS 25, 26. Before starting we require special contents in DL 1, 14 and DL 1, 15. There are

DL 1, 14   0000,0101,0000,0000,0100,0110,0000,0000
DL 1, 15   0000,0000,0000,0000,0000,0100,0000,0000

the expression in DL 1, 14 representing the order to transfer DL 3, 1 to TS 6.

〚47〛   *CALPOL 1.* Clear TS 22, 23; DL 1, 14–TS 27; DL 1, 15–TS 29. CALPOL 8.
CALPOL 8. B, BURY; B, INDEXIN; TS 28–TS 18; B, BURY; B, PLUSIND; B, BURY; B, INDEXIN; TS 28–TS 19; B, BURY; B, ADD; B, BURY; B, PLUSIND; TS 27–TS 2; TS 29–TS 3; AND; Q, CALPOL 40; TS 6–TS 15; Q, CALPOL 37; TS 6–TS 16; B, BURY; B, DISCRIM; B, 1.
CALPOL 37. TS 13–TS 18; TS 14–TS 19; B, BURY; B, TRANS 45; B, BURY; B, MULTIP; B, BURY; B, TRANS 45. CALPOL 49.
CALPOL 49. B, CALPOL 8.
CALPOL 50. TS 22–TS 25; TS 23–TS 26; B, UNBURY.

The above table for CALPOL has been expressed in a more abbreviated form than the one we gave for INDEXIN, several operations being listed at a time. AND is of course the logical operation and B, 1 indicates B with a 1 in column 17.

Outline of operation (CALPOL)   If we denote the polynomial by $a_1 x^{15} + a_2 x^{14} + \cdots$ the calculation proceeds by the equations $b_1 = a_1$, $c_1 = b_1 x$, $b_2 = c_1 + a_2$, $c_2 = b_2 x, \ldots$ After the calculation of each b_r we have 〚48〛 to determine whether this is the one required, viz. b_{16} or not. This is done by examining the content of TS 27 which includes the number r and is also, one might say principally, used to describe the position of the next coefficient a_{r+1}. If it is the one required we find ourselves at CALPOL 40 and have to pass b_r out to TS 25, 26. Otherwise we go to CALPOL 31, and after multiplying b_r by x to give c_r we find ourselves back at CALPOL 8 and repeating processes we have done before.

It will be evident that the table CALPOL is somewhat wasteful of space.   Each time a subsidiary operation is required we have to repeat B, BURY, and each time we make a transfer we have to do it in two stages, each of which uses a whole minor cycle of which most is wasted. It is possible to avoid this waste of space by keeping the instruction tables in some abbreviated form, and expanding each table whenever we want it. This will require a table EXPAND,   and will require each table to include appropriate references to the table EXPAND. These references will however be put in by EXPAND itself (when working under contract to a higher authority), just as EXPAND will put in the references to BURY and UNBURY.

BINDEC (general description)   The number in TS 13, 14 is translated into decimal form of the type $\alpha \times 10^m$ where $1 \leqslant \alpha < 10$, and is transferred into DL 10. The notation of the decimal form is such that the content of DL 10 can be passed out onto a card in the usual way and if the card is then listed the digits of the numbers α, m will then appear on the listing paper in the usual way. Or in other words only the first 10 minor cycles of DL 10 are used, and a decimal digit is represented by the minor cycle in which a pulse occurs, and its significance by the position of it within the minor cycle. (This account is incomplete as regards signs and some other details).

14.   The Design of Delay Lines

(i) *General.* A considerable amount of work has been done on delay lines for R.D.F. purposes.   On the whole our problems coincide with the R.D.F. problems but there are a few differences.

   (a) Owing to the fact that there will be more than one tank used in the calculator the stability of the delay is of importance. In R.D.F. the delay is allowed to determine the recurrence frequency and the effects of variations in it are thereby eliminated.

   (b) In R.D.F. it is required that the delayed signal should not differ from the undelayed by an error signal which is less than 60 dB (say) down on the signal proper. We are less difficult to please in this respect. We only require to be able to distinguish mark from space with a very high probability (e.g. at least $1 - 10^{-32}$). This requires a high signal to noise ratio, so far as the true random noise and the interference are concerned, but it does not require much as regards hum, frequency distortion and other factors producing unwanted signals of fairly constant amplitude.

[[60]]

Our main concerns then in designing a delay line will be:

(1) To ensure sufficient signal strength that noise does not cause serious effects.
(2) To eliminate or correct frequency and phase distortion sufficiently that we may correctly distinguish mark and space.
(3) To stabilise the delay to within say 0.2 pulse periods.
(4) To eliminate interference.
(5) To provide considerable storage capacity at small cost.
(6) To provide means for setting the crystals sufficiently nearly parallel.

The questions of noise and signal strength are treated in some detail in the following pages. It is found that there is plenty of power available unless either very long lines or very high frequencies are used. The elimination of interference is mainly a matter of shielding and is a very standard radio problem, which in our case is much less serious than usual. Various means have been found by the R.D.F. workers for setting the crystals. Some prefer to machine the whole delay line very accurately, others to provide means for moving the crystals through small angles, e.g. by bending the tank. All are satisfactory.

I list below a number of questions which must be answered in our design of delay lines. In order to fix ideas I have added the most probable answers in brackets after each question.

[49]   (1) What liquid should be used in the line? (Either mercury or a water-alcohol mixture).
(2) Should we use a carrier? If so, of what frequency? (Yes, certainly use a carrier. Frequency should be about 10 Mc/s with water-alcohol mixture, but may be higher if desired when mercury is used).
(3) What should be the clock-pulse frequency? (1 Mc/s).
(4) What should be the dimensions of the crystals? (Diameter might be half that of the tank, e.g. 1 cm. Thickness should be such that the first resonances of the two crystals are two or three megacycles on either side of the carrier, if water-alcohol is used. With mercury the thickness is less critical and may be either as with water-alcohol or may have resonance equal to carrier.
(5) Should the inside of the tank be rough or smooth? (Smooth).
(6) What should be the dimensions of the tank? (Standard tanks to give a delay of 1 ms. should be about 5' long whether water-alcohol or mercury. Diameter $\frac{1}{2}'$).

[61]

(Keep all the tanks within one degree Fahrenheit in temperature. Correct systematic temperature changes by altering the pulse frequency.)

In order to be able to answer these questions various mathematical problems connected with the delay lines will have to be solved.

(ii) *Electromagnetic conversion efficiency*. The delay line may best be considered as forming an electrical network of the kind usually (rather misleadingly) described as 'four-pole', i.e. a network which has one input current and one input voltage which together determine an output voltage and current. Such a network is described by three complex numbers at each frequency. In the case where there is little coupling between the output and input, which will apply to our problem, we may take these quantities to be the input and output admittances and the 'transfer admittance'. Strictly speaking we should specify whether the output is open circuit or short circuit when stating the input impedance, but with weak coupling these are effectively the same; similarly for the output impedance. The transfer admittance is the current produced at one end due to unit voltage at the other, and does not depend on which end has the voltage applied to it. In the case of the delay lines the input and output admittances will be effectively the capacities between the crystal electrodes. We need only determine the transfer admittance.

We shall consider the following idealised case. Two crystals of thickness d and d' are immersed in a liquid, with their faces perpendicular to the x-axis. The liquid extends to infinity in both the positive and the negative x-directions, and both liquid and crystals extend to infinity in the y and z directions (Fig. 40). The distance between the near side faces of the crystals is l. It is assumed that there is considerable attenuation of sound waves over a distance of the order of l but hardly any over a distance of the order of d or d'.

These assumptions are introduced largely with a view to eliminating the possibility of reflections. In practice the reflections would be eliminated by other means. For instance, the infinite liquid on the extreme right and left would be replaced by a short length of liquid in a stub of not very regular shape, so that the reflected waves would not be parallel to the face of the crystal. More likely still, of course, we should have some entirely different medium there.

The physical quantities involved are:

(a) The density ρ. We write ρ for the density of the crystal and ρ_1 for that of the liquid. Likewise a suffix 1 will indicate liquid values throughout.

[[62]]

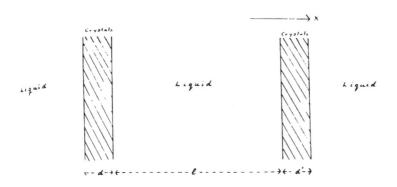

FIG.40   IDEAL ARRANGEMENT OF CRYSTALS

(b) The pressure p. In the case of the crystal this is understood to mean the xx-component of stress.

(c) The displacement ξ in the x-direction.

(d) The velocity v in the x-direction.

(e) The radian frequency ω.

(f) The elasticity η. This is the rate of change of pressure per unit decrease of logarithm of volume due to compression.

(g) The velocity of propagation c.

(h) The mechanical characteristic impedance ζ.

(i) The reciprocal radian wave length β.

(j) The piezo-electric constant ε. This gives the induced pressure due to an electric field strength of unity. This field strength should normally be thought of as in the x-direction, but we shall have to consider the case of a field in the y or z direction briefly also.

These quantities are related by the equations

$$c = \sqrt{\eta/\rho}, \quad \zeta = \sqrt{\eta\rho}, \quad \beta = \frac{\omega}{c}, \quad v = i\omega\xi, \quad i\omega\rho v = -\frac{dp}{dx},$$

$$p = -\eta\frac{d\xi}{dx} + E\varepsilon$$

In what follows we assume that all quantities such as p, v, ξ depend on time according to a factor $e^{i\omega t}$, which we omit.

[63]

We now consider the 'transmitting crystal', which we suppose extends from $x = -a$ to $x = a$ where $d = 2a$. The solution of the equations will be of form

$$p = E\varepsilon + B\cos\beta x$$

within the crystal, i.e. for $|x| < a$. Since the pressure is continuous we shall have

$$p = (E\varepsilon + B\cos\beta a)e^{i\beta_1(a-|x|)} \quad \text{if } |x| > a.$$

This gives for the velocity

$$v = \frac{1}{\omega\rho}\cdot -B\beta\sin\beta x = -iB\zeta^{-1}\sin\beta x \quad \text{if } |x| < a$$

$$v = \zeta_1^{-1}(E\varepsilon + B\cos\beta a)e^{i\beta_1(a-|x|)}\operatorname{sgn} x \quad \text{if } |x| > a.$$

Continuity of velocity now gives

$$B\left(\cos\beta a + \frac{i\zeta_1}{\zeta}\sin\beta a\right) = -E\varepsilon$$

and therefore the velocity at a is

$$\frac{-iB\sin\beta a}{\zeta} = \frac{iE\varepsilon\sin\beta a}{\zeta\cos\beta a + i\zeta_1\sin\beta a}$$

i.e. the velocity at the inside edge of the crystal is

$$\frac{iE\varepsilon}{\zeta}\cdot\frac{1}{\cot(d\omega/2c) + iu}$$

where $u = \zeta_1/\zeta$.

Assuming that the exciting voltage is longitudinal we may say that

$$\frac{\text{Velocity}}{\text{Exciting voltage}} = \frac{i\varepsilon}{\zeta d}\cdot\frac{1}{\cot(d\omega/2c) + iu}.$$

The effect of the medium between the two crystals we will not consider just yet. Let us simply assume that

$$\frac{\text{Velocity at inside edge of receiving crystal}}{\text{Velocity at inside edge of transmitting crystal}} = \vartheta.$$

[64]

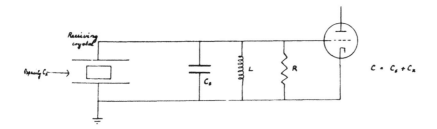

FIG. 41.   RECEIVING CRYSTAL CIRCUIT

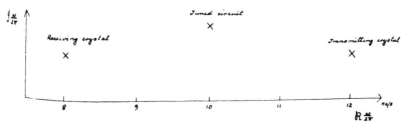

FIG. 42. SUGGESTED ARRANGEMENT OF POLES OF R(ω)

We have now to consider the effect of the receiving crystal. Fortunately we can deal with this by the principle of reciprocity. When applied to a mixed electrical and mechanical system this states that the velocity produced at the mechanical end by unit voltage at the electrical end is equal to the current produced at the electrical end by unit force at the mechanical end. Hence

$$\frac{\text{Current at receiving end}}{\text{Force on receiving crystal}} = \frac{i\varepsilon}{d'\zeta}\frac{1}{\cot(d'\omega/2c) + iu}.$$

To these equations we may add that the ratio of force to pressure is the area A' of the receiving crystal, and that the ratio of pressure to velocity is the mechanical characteristic impedance ζ_1. Combining we obtain

$$Y = \text{Transfer admittance} = \vartheta\frac{A'\varepsilon^2\zeta_1}{dd'\zeta^2}\frac{1}{(\cot(d\omega/2c) + iu)(\cot(d'\omega/2c) + iu)}.$$

Let us now assume that the input to the valve from the receiving crystal consists of a tuned circuit with a fairly low 'Q' as in Fig. 41. Then

Voltage attenuation and phase change factor

$$= \frac{\text{Grid voltage}}{\text{Input voltage}}$$

$$= \frac{Y}{(1/Li\omega) + Ci\omega + (1/R)}$$

$$= \frac{Y}{Ci\omega_0} \frac{\omega\omega_0}{(\omega + \omega_0 + (i\omega_0/2Q))(\omega - \omega_0 + (i\omega_0/2Q))}$$

where

$$LC\omega_0{}^2 l + \frac{1}{4Q^2} = 1, \quad C = C_s + C_x$$

$$Q = RC\omega_0$$

$$= \vartheta \frac{C_x}{C_x + C_s} \cdot \frac{2\pi\varepsilon^2}{\kappa\eta} \cdot R(\omega)$$

where

κ = Dielectric constant of crystal

ϑ = Attenuation due to viscosity of medium and geometrical causes.

$$R(\omega) = \frac{u}{(d\omega_0/2c)(\cot(d\omega/2c) + iu)(\cot(d'\omega/2c) + iu)}$$

$$\cdot \frac{\omega\omega_0}{(\omega + \omega_0 + (i\omega_0/2Q))(\omega - \omega_0 + (i\omega_0/2Q))}.$$

The quantity $2\pi\varepsilon^2/\kappa\eta$ depends only on the crystal, i.e. on the material of which it is made and its cut and form of excitation. Both ε^2 and η are of the dimensions of a pressure. $4\pi\varepsilon$ is of the dimensions of an electric field, and may be thought of as a constant electric field which has to be added to the varying field in order that the combination should produce the correct pressure variations, somewhat like the permanent magnet field in a telephone receiver. A typical value for $2\pi\varepsilon^2/\kappa\eta$ is 0.004.

Let us now consider the frequency-dependent factor, $R(\omega)$. The parameter u entering here is the ratio of the characteristic impedances of the

[66]

crystal and the liquid. It is equal to

$$\frac{\text{Velocity of sound in liquid} \times \text{density of liquid}}{\text{Velocity of sound in crystal} \times \text{density of crystal}}.$$

The velocity of sound in the crystal (X-cut quartz) is 5.72 km/sec. and its density is 2.7. The velocity in water is 1.44 km/sec., and the density 1, hence

$u(\text{water}) = 0.1$ abt.

The velocity in mercury is much the same but the density is 13.5. Hence

$u(\text{mercury}) = 1.3$ abt.

These figures suggest that we consider the two cases where u is small and where u is 1. The latter case may be described by saying that the liquid matches the crystal.

It may be assumed for the moment that our object is to make the minimum value of $|R(\omega)|$ in a certain given band of frequencies as large as possible. If the width of the band is 2Ω and it is centred on ω_0 and if we ignore the variations in ϑ we shall find that the optimum value of u is of the form $N\Omega/\omega_0$ where N is some numerical constant probably not too far from 1. The value of Q should be as large as possible. With $\Omega = 1$ Mc/s, $\omega_0 = 10$ Mc/s this seems to suggest that water ($u = 0.1$) is very suitable. In practice the differences due to the value of ϑ are more serious than those due to u, and there is in any case plenty of power. We would not in practice take Q as large as we could but would rather try to arrange that $|R(\omega)|$ was fairly constant throughout the band concerned and arg $R(\omega)$ fairly linear when plotted against ω. If water were used one would probably choose the thicknesses of the crystals and the value of Q to give poles of $|R(\omega)|$ somewhat as shown in Fig. 41. With this arrangement of the poles the gain corresponding to $|R(\omega)|$ is 9 dB throughout the range 8 Mc/s and the phase characteristic lies within 5° of the straight line within this range.

With mercury where u is nearly 1 we should put

$$\frac{d\omega_0}{2c} = \frac{\pi}{2}, \frac{d'\omega_0}{2c} = \frac{\pi}{2},$$

and then

$$|R(\omega)| = \frac{2}{\pi}\left(\sin\frac{\pi}{2}\frac{\omega}{\omega_0}\right)^2 \left| \frac{\omega\omega_0}{(\omega + \omega_0 + (i\omega_0/2Q))(\omega - \omega_0 + (i\omega_0/2Q))} \right|.$$

We should probably find it desirable to omit the tuned circuit, in which case $R(\omega)$ would represent a fairly constant loss of 4 dB. One could use a Q of 2 if one wished, giving a gain of 2 dB instead.

We have assumed above that the crystal is longitudinally excited. If it were transversely excited the figures would be much less satisfactory. At the transmitting end a far larger voltage would have to be applied in order to obtain the same field strength, and at the receiving end the stray capacities will have a more serious effect with transverse electrodes, although if the stray capacity were zero transverse electrodes at the receiving end would actually be more efficient.

(iii) *Geometrical attenuation.* If a rectangular crystal is crookedly placed in a plane parallel beam, the tilt being such that the one edge of the crystal is advanced in phase by an angle ψ then the attenuation due to the tilt is $\sin\frac{1}{2}\psi/\frac{1}{2}\psi$. With a square crystal whose side is 1 cm. and a frequency of 15 Mc/s this would mean that we get the first zero in the response for a tilt of about 16′. The setting is probably not really as critical as this owing to curvature of the wave fronts. If the crystals are operating in a free medium without the tube this effect is easily estimable and we find that, for crystals sufficiently far apart the allowable angles of tilt are of the order of the angle subtended at one crystal by the other. It has been found experimentally with tubes operating at 15 Mc/s that tilts of the order of half a degree are admissible.

Now let us consider the loss due to boundary effects. We assume a wave inside the tank of form $p = J_0(\beta'r)e^{-i\beta z + i\omega t}$ and assume a boundary condition of form $(1/p)(dp/dn) = \zeta$ where we do not know ζ nor even whether it is real or complex. The radius of the tank is a, so that the boundary condition becomes $\beta'aJ_1(\beta'a)/J_0(\beta'a) = \zeta a$. Let the solution of $uJ_1(u)/J_0(u) = y$ be $u(y)$. Then we have $\beta^2 + (u(\zeta a)/a)^2 = \omega^2/c^2$ and therefore $R\beta J\beta + (1/a^2)RuJu = 0$. But since $u(\zeta a)/\beta a$ is small this means approximately $J\beta = cRuJu/a^2\omega$, and the loss in a length l of the tank is $(lc/a^2\omega)RuJu$ nepers. For a given value of ζ there are many solutions of $uJ_1/J_0 = \zeta a$ but there is a bounded region of the u plane in which there is always a solution whatever value ζa may have. This means to say that for any boundary condition there is always a mode in which the attenuation does not exceed $\tau(c/a^2\omega)$ where τ is some numerical constant.

The value of τ is about 1.9. It is the largest value of xy such that $(x + iy)J_1(x + iy)/J_0(x + iy)$ is pure imaginary and $y > 0$, $0 < x < 2.4$.

[[68]]

Taking $lc/a^2\omega_0 = 0.31$ (as p. 90) the maximum loss in this mode is 6 dB. We should however probably add a certain amount to allow for the fact that not all of the energy will be in this mode. A total loss of 10 dB would probably not be too small.

(iv) *Attenuation in the medium.* The attenuation coefficient is given by $2\omega^2 v/3c^3$ where v is the dynamic coefficient of viscosity, i.e., the ratio of viscosity to density. With water ($v = .013$ $c = 1.44$ Km/sec.) at a frequency of 10 megacycles and a delay of 1 ms we have a loss of 12 dB. With mercury under the same circumstances the loss is only 1 dB.

These figures suggest that if water is used the frequency should not be much above 10 Mc/s, but that we can go considerably higher with mercury.

(v) *Noise.* Before leaving the subject of attenuation we should verify how much can be tolerated. The limiting factor is the noise, due to thermal agitation and to shot effect in the first amplifying valve. The effect of these is equivalent to an unwanted signal on the grid of the first valve, whose component in a narrow band of width f cycles has an R.M.S. value of

$$V_N = 4kTf(R + R_e)$$

where T is the absolute temperature, k is Boltzmann's constant and R is the resistive component of the impedance of the circuit working into the first valve, including the valve capacities. R_e is a constant for the valve and describes the shot effect for the valve. In the case that we use mercury and do not tune the input the value of R will be quite negligible in comparison with R_e, which might typically be 1000 ohms. For a pulse frequency of 1 megacycle we must take $f = 10^6$ (the theoretical figure is $\frac{1}{2}10^6$ but this is only attainable with rather peculiar circuits). At normal temperatures $4kT = 1.6 \times 10^{-20}$ and therefore $V_N = 4\,\mu\text{V}$. In the case that we use water and tune the input, we have $R = Q/\omega(C_x + C_s)$ at the worst frequency. Assuming $\omega/2\pi Q = 2$ Mc/s (see Fig. 41) and $C_x + C_s = 20$ pf and ignoring the fact that the effect will not be so bad at other frequencies, we have $V_N = 9\,\mu\text{V}$.

Now suppose that we wish to make sure that the probability of error is less than p, and that the difference in signal voltage between a digit 0 and a digit 1 is V. Then we shall need

$$2\int_{V/2V_N}^{\infty} e^{-\frac{1}{2}x^2}\,dx < p.$$

(This follows from the fact that a random noise voltage is normally distributed in all its coordinates). If we put $p = 10^{-32}$ we find

$$\frac{V}{V_N} \geqslant 24, \quad V \geqslant 0.1 \text{ mV}.$$

(vi) *Summary of output power results.* Summarising the voltage attenuation and noise questions we have:

(a) There is an attenuation factor depending on the material of the crystal and its cut and for quartz typically giving a loss of 48 dB.

(b) There is a factor R depending on the ratio of band width required to carrier frequency, and the matching factor u between crystal and liquid. In practical cases this amounts to gains of 10 dB with water and 2 dB with mercury.

(c) There is a loss factor $C_x/C_x + C_s$ due to stray capacity C_s across the receiving crystal. This might represent a loss of 6 dB.

(d) There is a loss due to the viscosity of the medium. For a water tank with a delay of 1 ms. and a carrier of 10 Mc/s the loss may be 12 dB: with mercury and a carrier of 20 Mc/s it may be 4 dB.

(e) Losses in the walls of the tank. Apparently this should not exceed 10 dB.

(f) The noise voltage may be 4×10^{-6} volts RMS (mercury) or 9×10^{-6} volts RMS (water).

(g) The signal voltage (peak to peak) should exceed the noise voltage (RMS) by a factor of 24 for safety.

These figures require input voltages (peak to peak) of 0.2 volts or 4.5 volts with mercury and water respectively. We could quite conveniently put 200 volts on, so that we have 60 dB (or 53 dB) to spare. There is no danger of breaking the crystals when they are operated with so much damping.

(vii) *Phase distortion due to reflections from the walls.* We cannot easily treat this problem quantitatively because of lack of information about the boundary conditions and because the ratio of diameter of crystal to diameter of tank is significant. Let us however try to estimate the order of magnitude by assuming the pressure zero on the boundary and considering the gravest mode. In this case the pressure is of form $J_0(k_1 r/a)e^{-i\beta z + i\omega t}$ where $2a$ is the diameter of the tank and $k_1 = 2.4$ is the smallest zero of J_0, and $\beta^2 + (k_1^2/a^2) = \omega^2/c^2$. In this case the change of phase down the length l of tank is $\varphi = l(\omega^2/c^2) - (k_1^2/a^2)$. If we are using carrier working we are chiefly interested in $d^2\varphi/d\omega^2$ which turns out to be $-k_1^2 cl/\omega_0^3 a^2$

[70]

where ω_0 is the carrier frequency. If we suppose that the band width involved is 2Ω, then the greatest phase error which is introduced is $k_1^2\Omega^2 cl/2\omega_0^3 a^3$. Let us suppose that the greatest admissible error is 0.2 radians, then we must have

$$\frac{lc}{a^2\omega_0} \leqslant \frac{0.4}{k_1^2}\left(\frac{\omega_0}{\Omega}\right)^2.$$

Taking

$$\omega_0 = 10 \text{ Mc/s}$$

$$\Omega = 1 \text{ Mc/s}$$

$$c = 1.4 \times 10^5 \text{ cm/sec.}$$

$$l = 1.4 \times 10^2 \text{ cm.}$$

$$a = 1 \text{ cm.}$$

Then

$$\frac{c}{\omega_0} = 2.2 \times 10^{-3} \text{ cm.}$$

$$\frac{lc}{a^2\omega_0} = 0.31$$

$$\frac{0.4}{k_1^2}\left(\frac{\omega_0}{\Omega}\right)^2 = 6.95$$

The situation is thus entirely satisfactory. The carrier frequency could even be halved.

(viii) *The choice of medium.* In choosing the medium we have to take into account

(a) That a medium with a small characteristic impedance such as water has a slight advantage as regards the factor $R(\omega)$.
(b) That water is more attenuative than mercury.
(c) That mercury gives wide band widths more easily than water because of closer matching, but that adequate band widths are nevertheless possible with water.
(d) That a water-alcohol mixture can be made to have a zero temperature coefficient of velocity at ordinary temperatures.

[71]

On the whole the advantages seem to be slightly on the side of mercury.

(ix) *Long lines*. The idea of using delay lines with a long delay, e.g. of the order of 0.1 second, is attractive because of the very large storage capacity that such a line would have. Although the long delay would make these unsuitable for general purposes they would be very suitable for cases where very large amounts of information were to be stored: in the majority of such cases the material is used in a fairly definite order and the long delay does not matter.

However such long lines do not really seem to be very hopeful. In order to reduce the attenuation to reasonable proportions it would be necessary to abandon carrier working, or else to use mercury. In either case we should probably be obliged to make the tank in the form of a bath rather than a tube; in the former case in order to avoid the phase distortion arising from reflections from the walls, and in the latter to economise mercury, using a system of mirrors in the bath. In any case the technique would involve much development work.

We propose therefore to use only tanks with a delay of 1 ms.

(x) *Choice of parameters*. Considerations affecting the carrier frequency are:

(a) The higher the carrier frequency the greater the possible band width.
(b) The difficulty of cutting thin crystals, somewhat modified by the absence of necessity of frequency stability.
(c) The attenuation at high frequencies of the sound wave in the liquid.
(d) The difficulty of setting the crystals up sufficiently nearly parallel if the wavelength is short.
(e) The difficulty of amplification at high frequencies.

Of these (a) and (c) are the most important. A reasonable arrangement seems to be to choose a frequency at which the attenuation in the medium is about 15db.

With the comparatively low frequencies and with wide tanks the setting up difficulty will not be serious. With long lines we should probably not attempt to do temperature correction, but would rephase the output.

Considerations affecting the pulse frequency are:

(a) The limitation of the pulse frequency to a comparatively small fraction of the carrier frequency if water is the transmission medium, and the limitation of this carrier frequency.
(b) The finite reaction times of the valves.

[[72]]

(c) The greater capacity of a line if the frequency is high.

(d) Greater speed of operation of the whole machine if the pulse frequency is high.

(e) Cowardly and irrational doubts as to the feasibility of high frequency working.

If we can ignore (e) the other considerations appear to point to a pulse frequency of about 3 megacycles or even higher. We are however somewhat alarmed by the prospect of even working at 1 megacycle since the difficulty (b) might turn out to be more serious than anticipated.

Considerations affecting the diameter of the tank are:

(a) That the crystals are most conveniently adjusted to be parallel by bending the tanks and that the diameter should therefore not be too large.

(b) That the diameter should be at least large enough to accommodate the crystal.

(c) That small diameters give phase distortion (p. 89).

(d) That with mercury small diameters are economical. At a price of £1 sterling per 1 lb. avoirdupois of mercury a 1 ms. tank of diameter 1″ would contain mercury to the value of about £2-2-6.

A diameter of 1″ or rather less is usual in R.D.F. tanks and appears reasonable in view of these conditions.

(xi) *Temperature control system.* The temperature coefficient of the velocity of propagation in mercury is quite small at 15 Mc/s, being only 0.0003/degree centigrade. This means that if the length of a 1 ms. line is to be correct to within 0.2 ms. then the temperature must be correct to within two-thirds of a degree centigrade.

15. The Design of Valve-Elements

(i) *Outline of the problem.* To design valve-elements with properties as described in § 5 and to work at a frequency of say 30 or 100 kilocycles would be very straightforward. When the pulse recurrence frequency is as high as a megacycle we shall have to be more careful about the design, but we need not fear any real difficulties of principle about working at these frequencies, and with such band widths. The successful working of television equipment gives us every encouragement in this respect. A word of warning might perhaps be in order at this point. One is tempted to try and carry the argument further and try to infer something from the success of R.D.F. at frequencies of several thousands of megacycles. Such an analogy would however not be in order for although these very high frequencies are used the bandwidth of intelligence which can be transmitted is still comparatively small, and it is not easy to see how the band width could be greatly increased.

In this chapter I shall discuss the limitations inherent in the problem, and shall also show very tentative circuit diagrams by way of illustration. These circuits have not yet been tried out, and I have too much experience of electronic circuits to believe that they will work well just as they stand. (This does not represent a superstitious belief in the cussedness of circuits and the inapplicability of mathematics thereto. Rather it means that normally the amount of mathematical argument required to get a reliable prophecy of the behaviour of a circuit is out of proportion to the small trouble required to try it out, at any rate if one is in an electrical laboratory. In practice one compromises with a rough mathematical argument and then follows up with experiment. The apparent "cussedness" of electronic circuits is due to the fact that it is necessary to make rather a lot of simplifying assumptions in these arguments, and that one is very liable to make the wrong ones, by false analogy with other circuits one has dealt with on previous occasions. The cussedness lies more in the minds dealing with the problem than in the electronic circuits themselves.)

(ii) *Sources of delay*. There are two main reasons why vacuum tubes should cause delays, viz. the input capacity and the transit time. Of these perhaps the first is in practice the more serious, the second the more theoretically unavoidable.

The delay due to the input capacity, when the valves are driven to saturation or some other limiting arrangement is used, is of the order of C/g_m, where C is the input capacity and g_m is the mutual conductance of the valve. We may, for instance consider the idealised circuit Fig. 44. (Coupling with a battery is of course not practical politics, but it produces essentially the same effects as more practical circuits, and is more easily understood). If I is the saturation current then the grid swing required to produce it is I/g_m and the charge which must flow into the grid to produce this voltage is CI/g_m. If the whole saturation current is available the time required is C/g_m. This argument is only approximate, and omits some small purely numerical factors. However it illustrates the more important points. In particular we can see that Miller effect is not a very serious matter because of the limiting, which reduces the effective amplification factor to 1. On the other hand, if one valve is used to serve several inputs the delay will be correspondingly

[[74]]

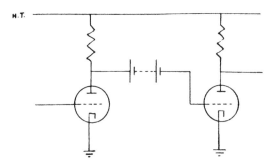

FIG. 44.

increased because the capacity has become multiplied by the number of grids served.

This connecting of several grids to one anode, and a number of other practical points will tend to make the actual delay due to input capacity several times greater than C/g_m, e.g. $10C/g_m$.

The delay due to transit time may be calculated, in the case of a plane structure, to be $3d(m/2eV)^{1/2}$ where m, e are respectively the mass and charge of the electron, V is the voltage of the grid referred to cut-off and d is the grid-cathode spacing. In other words the transit time may be calculated on the assumption that the average velocity of the electrons between cathode and grid is one-third of the velocity when passing the grid. This time may be compared with C/g_m which, if C is calculated statically, has the value $(3/2)d(m/2eV)^{1/2}$, i.e. half of the transit time. That there should be some such relation between C/g_m and transit time can be seen by calculating $C/(g \times \text{Transit time})$, where C is the grid-cathode capacity and g is the actual conductance, i.e., the ratio of current to V.

$$\frac{C}{g \times \text{Transit time}} = \frac{CV}{I \times \text{Transit time}}$$

$$= \frac{\text{Charge on grid}}{\text{Charge in transit}}.$$

Let us now calculate actual values. The voltage V by which the grid exceeds cut-off might be 10 volts which corresponds to a velocity about 1/300 of velocity of light (Note: annihilation energy of electron is half a million volts) or one metre per microsecond. If d is 0.2 cm. the transit time is 0.006 μs. A typical value for C/g_m is 0.002 μs.

The relation between C/g_m and transit time brings up an important point, viz. that these two phenomena of time delay are really inseparable. The input capacity of the tube when 'hot' really consists largely of a capacity to the electrons. When the motion of the electrons is taken into account the capacity is found to become largely resistive (Ferris effect).

Before proceeding further I should try to explain the way I am using the word 'delay'. When I say that there is a delay of so many microseconds in a circuit I do not mean to say that the output differs from the input only in appearing that much later. I wish I did. What I mean is something much less definite, and also less agreeable. Strictly speaking I should specify very much more than a single time. I should specify the waveform of the output for every input waveform, and even this would be incomplete unless it referred both to voltages and currents. We have not space to consider these questions, nor is it really necessary. I should however give some idea of what kind of distortion of output these 'delays' really involve. In the case of the input capacity the distortion may be taken to be of the form that an ideal input pulse of unit area is converted into a pulse of unit area with sharp leading edge and exponentially decaying trailing edge, the time constant of the decay being the 'delay', thus Fig. 44a. In the case of the transit time the curve is probably more nearly of the 'ideal' form (Fig. 44b).

To give the word 'delay' a definite meaning, at any rate for networks, I shall understand it to mean the delay for low frequency sine waves. This is equal to the displacement in time of the centre of gravity in the case of pulses.

In order to give an idea of the effect of these delays we have shown in Fig. 45 a pulse of width 0.2 μs and the same pulse delayed, after the manner of Fig. 44a, by 0.03 μs, this representing our calculated value of 0.003 multiplied by 10 to allow for numerous grids, etc. etc.   It will be seen that the effect is by no means to be ignored, but nevertheless of a controllable magnitude.

(iii) *Use of cathode followers.* In order to try and separate stages from one another as far as possible we shall make considerable use of cathode followers. This is a form of circuit which gives no amplification, and indeed a small attenuation (e.g. 0.5 dB); but has a very large input impedance and a very low output impedance. This means chiefly that we can load a valve with many connections into cathode followers without its output being seriously affected.

[76]

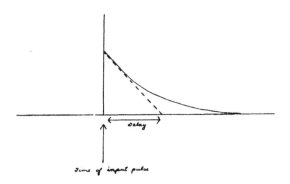

FIG. 44a

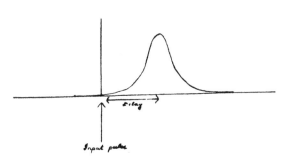

FIG. 44b

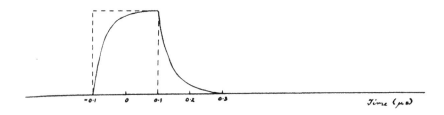

FIG. 45. PROBABLE EFFECT OF INPUT CAPACITY ON A SQUARE PULSE

SUCH AS THE CLOCK PULSE

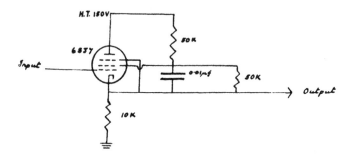

FIG. 46

Fig. 46 shows a design of cathode follower in which the input capacity effect has been reduced by arranging that the anode is screened from the grid and that the screen voltage as well as that of the cathode moves with the grid. If one could ignore transit time effects this would have virtually zero input capacity.

(iv) *The 'limiting amplifier' circuit*. When low frequencies are used the limiter circuit can conveniently be nothing more nor less than an amplifier, the limiting effect appearing at cut-off and when grid and cathode voltages are equal. At high frequencies we cannot get a very effective limiting effect at cathode voltage, owing to the fact that the grid must be supplied from a comparatively low impedance source to avoid a large delay arising from input capacity, but on the other hand, in order to get a limiting effect we need a high impedance, high compared with the grid conduction impedance (about 2000 ohms probably).

At high frequencies it is probably better to use a 'Kipp relay' circuit. This is nothing more than a multivibrator in which one leg has been made infinitely long (and then some), i.e. one of the two semi-stable states has been made really stable.   An impulse will however make the system occupy the other state for a time and then return, producing a pulse during the period in which it occupies the less stable state. This pulse can be taken in either polarity. It is fairly square in shape and its amplitude is sensibly independent of the amplitude of the tripping pulse, although its time may depend on it slightly. These are all definite advantages.

A suggested circuit is shown in Fig. 47, and the waveforms associated with it at various points in Fig. 48.

(v) *Trigger circuit*. The trigger circuit need only differ very little from the

A. M. Turing

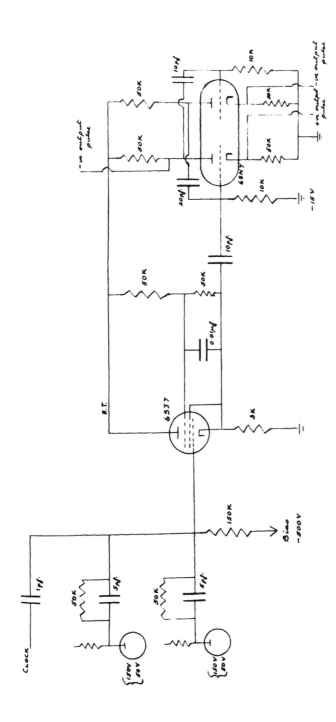

FIG. 47. 'LIMITING AMPLIFIER' CIRCUIT

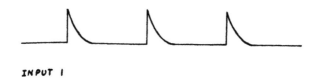

INPUT 1

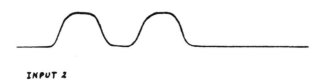

INPUT 2

INPUT TO CATHODE FOLLOWER

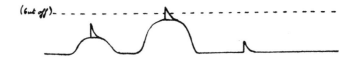

FIG. 48.

[[80]]

limiter or Kipp relay. It needs to have two quite stable states, and we therefore return both of the grids of the 6SN7 to −15 volts instead of returning one to ground. Secondly the inhibitory connection is different. In the case of the limiter it simply consists of an opposing or negative voltage on the cathode follower; in the case of the trigger circuit it must trip the valve back, and therefore we need a second cathode follower input connected to the other grid of the 6SN7.

(vi) *Unit delay*. The essential part of the unit delay is a network, designed to work out of a low impedance and into a high one. The response at the output to a pulse at the input should preferably be of the form indicated in Fig. 50, i.e. there should be a maximum response at time 1 μs after the initiating pulse, and the response should be zero by a time 2 μs after it, and should remain there. It is particularly important that the response should be near to zero at the integral multiples of 1 μs after the initiating pulse (other than 1 μs after it).

A simple circuit to obtain this effect is shown in Fig. 51a. The response is shown in Fig. 51b. It differs from the ideal mainly in having its maximum too early. It can be improved at the expense of a less good zero at 2 μs by using less damping, i.e. reducing the 500 ohm resistor. It is also possible to obtain altogether better curves with more elaborate circuits.

The 1000 ohm resistors at input and output may of course be partly or wholly absorbed into the input and output circuits. Further the whole impedance scale may be altered at will.

The fact that the pulse has become greatly widened in passing through the delay network does not signify. It will only be used to gate a clock pulse or to assist in tripping a Kipp relay, and therefore will give rise to a properly shaped pulse again.

(vii) *Trigger limiter*. We can build up a trigger limiter out of the other elements, although we cannot replace it by such a combination in the circuit diagrams because we are not putting a legitimate form of input into all of them. The circuit is (Fig. 52).

The valve P is merely a frequency divider. It can be used to supply all the trigger limiters. The trigger circuit Q should be tripped by the combination of pulse from P and continuous input, and will itself trip R. The arrangement of two trigger circuits prevents any danger of half-pulse outputs, which we are most anxious to aviod. In order that there might be a half-pulse output the trigger circuit Q would have to remain near its unstable state of equilibrium for a period of time of 1 μs. In order that this may

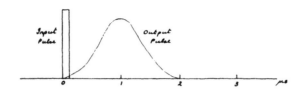

FIG. 50. INDICIAL RESPONSE DERIVATIVE FOR UNIT DELAY (preferable form)

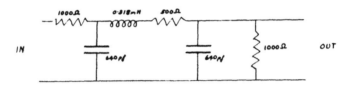

FIG. 51a

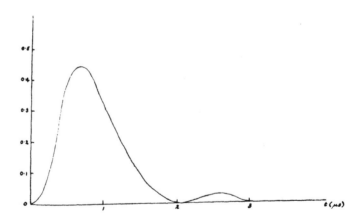

FIG. 51b A POSSIBLE UNIT DELAY CIRCUIT and CORRESPONDING INDICIAL RESPONSE
DERIVATIVE

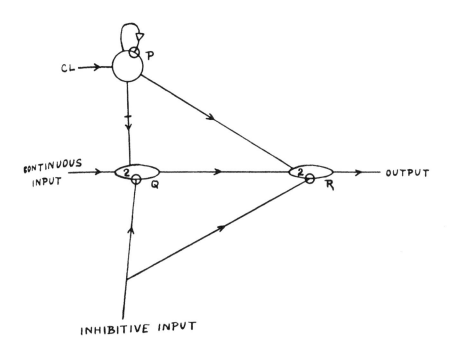

CL →

CONTINUOUS
INPUT →

OUTPUT

INHIBITIVE INPUT

FIG. 52

happen the magnitude of the continuous input voltage has to be exceed-ingly finely adjusted; the admissible range is of the form Ae^{-tg_mC} where A might be say 100 volts (it doesn't matter really) and t is the time between pulses, C and g_m the input capacity and mutual conductance of the valves used in the trigger circuit; C/g_m might be 0.002 μs (we do not need to allow for Miller effect), so that the admissible voltage range is about 10^{-200} volts which is adequately small.

16. Alternative Forms of Storage

(i) *Desiderate for storage systems.* A storage system should have a high *monetory economy*, i.e. we wish to be able to store a large number of digits per pound sterling of outlay: it should also have a high *spacial economy*. For the majority of purposes we like a form of storage to be *erasible*, although there are a number of purposes, such as function tables and the greater part of the instruction tables, for which this is not necessary. For the majority of purposes we also like to have a short *accessibility time*, defining the accessibility time to be the average time which one has to wait in order to find out the value of a stored digit. Normally we shall be interested in the values of a group of digits which are all stored close together, and very often it does not take much longer to obtain the information about the whole group than about the single digit. Let us say that the additional time necessary per digit required is the *digit time* (*reading*). We may also define the accessibility and digit times for recording in the obvious analogous way, though they are usually either equal to the reading time or else exceedingly long.

(ii) *Survey of available storage methods.* The accompanying table gives very rough figures for the various available types of storage and the quantities defined above. This table must not be taken too seriously. Many of the figures are based on definite numerical data, but most are guesses. In spite of the roughness of the figures the table brings out a number of points quite clearly.

(1) All the well established forms of storage (excepting the cerebral cortex) are either very expensive and bulky, or else have a very high accessibility time.

(2) The really economical systems consist of layers packed into the form of a solid. They are read by exposing the layer wanted.

(3) The systems which are both economical and fairly fast have the information arranged in two dimensions. This apparently applies even to the cerebral cortex.

(4) Much the most hopeful scheme, for economy combined with speed, seems to be the 'storage tube' or 'iconoscope' (in J. v. Neumann's terminology).

(5) Some use could probably also be made of magnetic tape and of film for cases where the accessibilty time is not very critical.

(iii) *Storage tubes.* In an iconoscope as used in television a picture of a scene is stored as a charge pattern on a mosaic, and is subsequently read by scanning the pattern with an electron beam. The electron beam brings the charge density back to a standard value and the charge lost by the mosaic registers itself through its capacity to a 'signal plate' behind the mosaic. The information stored in this way on an iconoscope, using a 500 line system, corresponds to a quarter of a million digits.

One might possibly use an actual iconoscope as a method of storage, but

[84]

A. M. Turing

	Monetary economy (digits/£)	Spacial economy (digits/litre)	Access time (reading)	Digit time (reading)	Access time (recording)	Digit time (recording)	Remarks
Inerasible systems.							
Punched paper tape	10^7	$5 \cdot 10^6$	$\geqslant$1 min.	.03 sec.	(= reading)	.03 sec.	Permutable
Hollerith cards	10^6	$3 \cdot 10^5$	$\geqslant$1 min.	1 ms.	(= reading)	1 ms.	Human use. Not very convenient for mechanical or electrical reading.
Print on paper	10^8	10^8	30 secs.	10 ms.			
Film (a) Displayed stationary	10^4	10^4	5 μs	1 μs		1 μs	
(b) Wound on reels	10^9	$3 \cdot 10^{10}$	$\geqslant$1 min.	1 μs		1 μs	
Soldered connections	1000	200	<1 μs	<1 μs	15 mins.	1 min.	
Erasible systems.							
Plugboards	50	50	<1 μs	<1 μs	30 secs.	10 secs.	
Wheels, etc.	20	$2 \cdot 10^3$	30 ms.	30 ms.	30 ms.	30 ms.	(Mechanically read).
Relays	2	2	<1 μs	<1 μs	10 ms.	10 ms.	
Thyratrons	2	2	<1 μs	<1 μs	10 μs	10 μs	
Neons	20	50	<1 μs	<1 μs	30 μs	30 μs	
Trigger circuits	3	3	<1 μs	<1 μs	1 μs	<1 μs	
Cerebral cortex	10^5	10^9	5 sec.	30 ms.	30 sec.	5 sec.	Man at £300 p.a. capitalised
Acoustic delay lines	200	50	1 ms.	1 μs	1 ms.	1 μs	More optimistic estimate than in §10. Circular wave guide with 1 cm. waves.
Electric delay lines	100	200	100 μs	<1 μs	100 μs	<1 μs	Numerous carriers.
Storage tubes	10^4	10^4	5 μs	1 μs	5 μs	1 μs	Described as 'Iconoscope' by J. v. Neumann.
Magnetic tape	10^8	$3 \cdot 10^8$	1 min.	10^{-4} sec.	1 min.	10^{-4} sec.	

there are better arrangements. Instead of putting the charge pattern on to the 'mosaic' with light we can put it on with an electron beam. The density of the charge pattern left by the beam can be varied by modulating either the voltage of the signal plate or the current in the beam. The advantages of this are:

(a) The charge pattern can be set up more quickly with an electron beam than with light.
(b) Less apparatus is required.
(c) The same beam can be used for reading and recording, so that distortion of the pattern does not matter.

It seems probable that a suitable storage system can be developed without involving any new types of tube, using in fact an ordinary cathode ray tube with tin-foil over the screen to act as a signal plate. It will be necessary to furbish up the charge pattern from time to time, as it will tend to become dissipated. The pattern is said to last for days when there is no electron beam, but if we have a beam scanning one part of the target it will send out secondary electrons which will tend to destroy the remainder of the pattern. If we were always scanning the pattern in a regular manner as in television this would raise no serious problems. As it is we shall have to provide fairly elaborate switching arrangements to be applied when we wish to take off a particular piece of information. It will be necessary to stop the beam from scanning in the refurbishing cycle, switch to the point from which the information required is to be taken, do some scanning there, replace the information removed by the scanning, and return to refurbishing from the point left off. Arrangements must also be made to make sure that refurbishing does not get neglected for too long because of more pressing duties. None of this involves any fundamental difficulty, but no doubt it will take time to develop.

⟦86⟧

3 Lecture to the London Mathematical Society on 20 February 1947

A. M. Turing

The automatic computing engine now being designed at N.P.L. is a typical large scale electronic digital computing machine. In a single lecture it will not be possible to give much technical detail of this machine, and most of what I shall say will apply equally to any other machine of this type now being planned.

From the point of view of the mathematician the property of being digital should be of greater interest than that of being electronic. That it is electronic is certainly important because these machines owe their high speed to this, and without the speed it is doubtful if financial support for their construction would be forthcoming. But this is virtually all that there is to be said on that subject. That the machine is digital however has more subtle significance. It means firstly that numbers are represented by sequences of digits which can be as long as one wishes. One can therefore work to any desired degree of accuracy. This accuracy is not obtained by more careful machining of parts, control of temperature variations, and such means, but by a slight increase in the amount of equipment in the machine. To double the number of significant figures used would involve increasing the equipment by a factor definitely less than two, and would also have some effect in increasing the time taken over each job. This is in sharp contrast with analogue machines, and continuous variable machines ⟦1⟧ such as the differential analyser, where each additional decimal digit required necessitates a complete redesign of the machine, and an increase in the cost by perhaps as much as a factor of 10. A second advantage of digital computing machines is that they are not restricted in their applications to any particular type of problem. The differential analyser is by far the most general type of analogue machine yet produced, but even it is comparatively limited in its scope. It can be made to deal with almost any kind of ordinary differential equation, but it is hardly able to deal with partial ⟦2⟧ differential equations at all, and certainly cannot manage large numbers of linear simultaneous equations, or the zeros of polynomials. With digital machines however it is almost literally true that they are able to tackle any computing problem. A good working rule is that the ACE can be made to do any job that could be done by a human computer, and will do it in one ten-thousandth of the time. This time estimate is fairly reliable, except in cases where the job is too trivial to be worth while giving to the ACE.

Some years ago I was researching on what might now be described as an ⟦3⟧ investigation of the theoretical possibilities and limitations of digital computing machines. I considered a type of machine which had a central mechanism, and an infinite memory which was contained on an infinite tape. This type of machine appeared to be sufficiently general. One of my conclusions was that the idea of a 'rule of thumb' process and a 'machine process' were synonymous. The expression 'machine process' of course means one which could be carried out by the type of machine I was considering. It was essential in these theoretical arguments that the memory should be infinite. It can easily be shown that otherwise the machine can only execute periodic operations. Machines such as the ACE may be regarded as practical versions of this same type of machine. There is at least a very close analogy. Digital computing machines all have a central mechanism or control and some very extensive form of memory. The memory does not have to be infinite, but it certainly needs to be very large. In general the arrangement of the memory on an infinite tape is unsatisfactory in a practical machine, because of the large amount of time which is liable to be spent in shifting up and down the tape to reach the point at which a particular piece of information required at the moment is stored. Thus a problem might easily need a storage of three million entries, and if each entry was equally likely to be the next required the average journey up the tape would be through a million entries, and this would be intolerable. One needs some form of memory with which any required entry can be reached at short notice. This difficulty presumably used to worry the Egyptians when their books were written on papyrus scrolls. It must have been slow work looking up references in them, and the present arrangement of written matter in books which can be opened at any point is greatly to be preferred.

[[4]]

We may say that storage on tape and papyrus scrolls is somewhat *inaccessible*. It takes a considerable time to find a given entry. Memory in book form is a good deal better, and is certainly highly suitable when it is to be read by the human eye. We could even imagine a computing machine that was made to work with a memory based on books. It would not be very easy but would be immensely preferable to the single long tape. Let us for the sake of argument suppose that the difficulties involved in using books as memory were overcome, that is to say that mechanical devices for finding the right book and opening it at the right page, etc. etc. had been developed, imitating the use of human hands and eyes. The information contained in the books would still be rather inaccessible because of the time occupied in the mechanical motions. One cannot turn a page over very quickly without tearing it, and if one were to do much transportation, and

[[5]]

do it fast, the energy involved would be very great. Thus if we moved one book every millisecond and each was moved ten metres and weighed 200 grams, and if the kinetic energy were wasted each time we should consume 10^{10} watts, about half the country's power consumption. If we are to have a really fast machine then, we must have our information, or at any rate a part of it, in a more accessible form than can be obtained with books. It seems that this can only be done at the expense of compactness and economy, e.g. by cutting the pages out of the books, and putting each one in to a separate reading mechanism. Some of the methods of storage which are being developed at the present time are not unlike this.

If one wishes to go to the extreme of accessibility in storage mechanisms one is liable to find that it is gained at the price of an intolerable loss of compactness and economy. For instance the most accessible known form of storage is that provided by the valve flip-flop or Jordan Eccles trigger   [6] circuit. This enables us to store one digit, capable of two values, and uses two thermionic valves. To store the content of an ordinary novel by such means would cost many millions of pounds. We clearly need some compromise method of storage which is more accessible than paper, film etc, but more economical in space and money than the straightforward use of valves. Another desirable feature is that it should be possible to record into the memory from within the computing machine, and this should be possible whether or not the storage already contains something, i.e. the storage should be *erasible*.

There are three main types of storage which have been developed recently and have these properties in greater or less degree. Magnetic wire is very compact, is erasible, can be recorded on from within the machine, and is moderately accessible. There is storage in the form of charge patterns on the screen of a cathode ray tube. This is probably the ultimate solution. It could eventually be nearly as accessible as the Jordan Eccles circuit. A third possibility is provided by acoustic delay lines. They give greater accessibility than the magnetic wire, though less than the C.R.T type. The accessibility is adequate for most purposes. Their chief advantage is that they are already a going concern. It is intended that the main memory of the ACE shall be provided by acoustic delay lines, consisting of mercury tanks.

The idea of using acoustic delay lines as memory units is due I believe to Eckert of Philadelphia University, who was the engineer chiefly responsible for the Eniac. The idea is to store the information in the form of com-

[[7]]

pression waves travelling along a column of mercury. Liquids and solids will transmit sound of surprisingly high frequency, and it is quite feasible to put as many as 1000 pulses into a single 5′ tube. The signals may be conveyed into the mercury by a piezo-electric crystal, and also detected at the far end by another quartz crystal. A train of pulses or the information

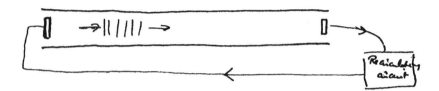

which they represent may be regarded as stored in the mercury whilst it is travelling through it. If the information is not required when the train emerges it can be fed back into the column again and again until such time as it *is* required. This requires a 'recirculating circuit' to read the signal as it emerges from the tank and amplify it and feed it in again. If this were done with a simple amplifier it is clear that the characteristics of both the tank and the amplifier would have to be extremely good to permit the signal to pass through even as many as ten times. Actually the recirculating circuit does something slightly different. What it does may perhaps be best expressed in terms of point set topology. Let the plane of the diagram represent the space of all possible signals. I do not of course wish to imply that this is two dimensional. Let the function f be defined for arguments in this signal space and have values in it. In fact let $f(s)$ represent the effect on the signal s when it is passed through the tank and the recirculating mechanism. We assume however that owing to thermal agitation the effect of recirculation may be to give any point within a circle of radius δ of $f(s)$. Then a necessary and sufficient condition that the tank can be used as a storage which will distinguish between N different signals is that there must be N sets $E_1 \ldots E_N$ such that if F_r is the set of points within distance ε of E_r

$$s \in F_r \supset f(s) \in E_r$$

and the sets F_r are disjoint. It is clearly sufficient for we have only then to ensure that the signals initially fed in belong to one or other of the sets F_r, and it will remain in the set after any number of recirculations, without any

[[90]]

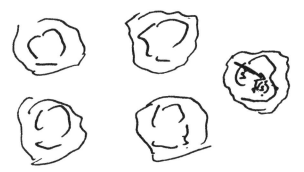

danger of confusion. It is necessary for suppose $s_1 \ldots s_N$ are signals which have different meanings and which can be fed into the machine at any time and read out later without fear of confusion.

Let E_r be the set of signals which *could* be obtained for s_r by successive applications of f and shifts of distance not more than ε. Then the sets E_r are disjoint [two lines indecipherable—Ed.]. In the case of a mercury delay line used for $N = 16$ the set would consist of all continuous signals within the shaded area.

One of the sets would consist of all continuous signals lying in the region below. It would represent the signal 1001.

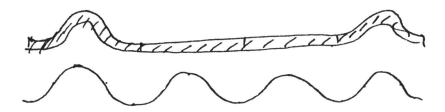

In order to put such a recirculation system into effect it is essential that a clock signal be supplied to the memory system so that it will be able to distinguish the times when a pulse if any should be present. It would for instance be natural to supply a timing sine wave as shown above to the recirculator.

The idea of a process f with the properties we have described is a very common one in connection with storage devices. It is known as 'regeneration' of storage. It is always present in some form, but sometimes the regeneration is as it were naturally occuring and no precautions have to be taken. In other cases special precautions have to be taken to improve such an f process or else the impression will fade.

The importance of a clock to the regeneration process in delay lines may be illustrated by an interesting little theorem. Suppose that instead of the condition $s \in F_r \supset f(s) \in E_r$ we impose a stronger one, viz $f''(s) \rightarrow c_r$ if $s \in E_r$, i.e. there are ideal forms of the distinguishable signals, and each admissible signal converges towards the ideal form after recirculating. Then we can show that unless there is a clock the ideal signals are all constants. For let U_α represent a shift of origin, i.e. $U_\alpha s(t) = s(t + \alpha)$. Then since there is no clock the properties of the recirculator are the same at all times and f therefore commutes with U_α. Then $fU_\alpha(c_r) = U_\alpha f(c_r) = U_\alpha c_r$, for $f(c_r) = c_r$ since c_r is an ideal signal. But this means that $U_\alpha(c_r)$ is an ideal signal, and therefore for sufficiently small α must be c_r, since the ideal signals are discrete. Then for any β and sufficiently large u, β/u will be sufficiently small and $U_{\beta/u}(c) = c$. But then by iteration $c = U_{\beta/u}^u(c) = U_\beta(c)$ i.e. $c(t + \beta) = c(t)$. This means that the ideal signal c is a constant.

We might say that the clock enables us to introduce a discreteness into time, so that time for some purposes can be regarded as a succession of instants instead of a continuous flow. A digital machine must essentially deal with discrete objects, and in the case of the ACE this is made possible by the use of a clock. All other digital computing machines except for human and other brains that I know of do the same. One can think up ways of avoiding it, but they are very awkward. I should mention that the use of the clock in the ACE is not confined to the recirculation process, but is used in almost every part.

It may be as well to mention some figures connected with the mercury delay line as we shall use it. We shall use five foot tubes, with an inside diameter of half an inch. Each of these will enable us to store 1024 binary digits. The unit I have used here to describe storage capacity is self explanatory. A storage mechanism has a capacity of m binary digits if it can remember any sequence of m digits each being a 0 or a 1. The storage capacity is also the logarithm to the base 2 of the number of different signals which can be remembered, i.e. $\log_2 N$. The digits will be placed at a time interval of one microsecond, so that the time taken for the waves to travel down the tube is just over a millisecond. The velocity is about one and a half kilometres per second. The delay in accessibility time or average waiting for a given piece of information is about half a millisecond. In practice this is reduced to an effective 150 μs. The full storage capacity of the ACE available on Hg delay lines will be about 200,000 binary digits. This is probably comparable with the memory capacity of a minnow.

I have spent a considerable time in this lecture on this question of memory, because I believe that the provision of proper storage is the key to the problem of the digital computer, and certainly if they are to be persuaded to show any sort of genuine intelligence much larger capacities than are yet available must be provided. In my opinion this problem of making a large memory available at reasonably short notice is much more important than that of doing operations such as multiplication at high speed. Speed is necessary if the machine is to work fast enough for the machine to be commercially valuable, but a large storage capacity is necessary if it is to be capable of anything more than rather trivial operations. The storage capacity is therefore the more fundamental requirement.   [[8]]

Let us now return to the analogy of the theoretical computing machines with an infinite tape. It can be shown that a single special machine of that type can be made to do the work of all. It could in fact be made to work as a model of any other machine. The special machine may be called the universal machine; it works in the following quite simple manner. When we have decided what machine we wish to imitate we punch a description of it on the tape of the universal machine. This description explains what the machine would do in every configuration in which it might find itself. The universal machine has only to keep looking at this description in order to find out what it should do at each stage. Thus the complexity of the machine to be imitated is concentrated in the tape and does not appear in the universal machine proper in any way.

If we take the properties of the universal machine in combination with the fact that machine processes and rule of thumb processes are synonymous we may say that the universal machine is one which, when supplied with the appropriate instructions, can be made to do any rule of thumb process. This feature is paralleled in digital computing machines such as the ACE. They are in fact practical versions of the universal machine. There is a certain central pool of electronic equipment, and a large memory. When any particular problem has to be handled the appropriate instructions for the computing process involved are stored in the memory of the ACE and it is then 'set up' for carrying out that process.

I have indicated the main strategic ideas behind digital computing machinery, and will now follow this account up with the very briefest description of the ACE. It may be divided for the sake of argument into the following parts

Memory
Control
Arithmetic part
Input and output

I have already said enough about the memory and will only repeat that in the ACE the memory will consist mainly of 200 mercury delay lines each holding 1024 binary digits. The purpose of the control is to take the right instructions from the memory, see what they mean, and arrange for them to ⟦9⟧ be carried out. It is understood that a certain 'code of instructions' has been laid down, whereby each 'word' or combination of say 32 binary digits ⟦10⟧ describes some particular operation. The circuit of the control is made in accordance with the code, so that the right effect is produced. To a large extent we have also allowed the circuit to determine the code, i.e. we have not just thought up an imaginary 'best code' and then found a circuit to put it into effect, but have often simplified the circuit at the expense of the code. It is also quite difficult to think about the code entirely *in abstracto* without any kind of circuit. The arithmetic part of the machine is the part concerned with addition, multiplication and any other operations which it seems worth while to do by means of special circuits rather than through the simple facilities provided by the control. The distinction between control and arithmetic part is a rather hazy one, but at any rate it is clear that the machine should at least have an adder and a multiplier, even if they turn out in the end to be part of the control. This is the point at which I should mention that the machine is operated in the binary scale, with two qualifications. Inputs from externally provided data are in decimal, and so are outputs intended for human eyes rather than for later reconsumption by the ACE. This is the first qualification. The second is that, in spite of the intention of binary working there can be no bar on decimal working of a kind, because of the relation of the ACE to the universal machine. Binary

working is the most natural thing to do with any large scale computer. It is   〚11〛
much easier to work in the scale of two than any other, because it is so easy
to produce mechanisms which have two positions of stability: the two
positions may then be regarded as representing 0 and 1. Examples are lever
as diagram, Jordan Eccles circuit, thyratron. If one is concerned with a

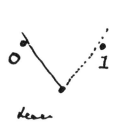

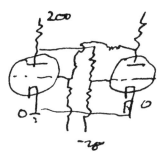

small scale calculating machine then there is at least one serious objection
to binary working. For practical use it will be necessary to build a converter
to transform numbers from the binary form to the decimal and back. This
may well be a larger undertaking than the binary calculator. With the large
scale machines this argument carries no weight. In the first place a converter would become a relatively small piece of apparatus, and in the second it
would not really be necessary. This last statement sounds quite paradoxical, but it is a simple consequence of the fact that these machines can be
made to do any rule of thumb process by remembering suitable instructions. In particular it can be made to do binary decimal conversion. For
example in the case of the ACE the provision of the converter involves no
more than adding two extra delay lines to the memory. This situation is
very typical of what happens with the ACE. There are many fussy little
details which have to be taken care of, and which, according to normal
engineering practice would require special circuits. We are able to deal with
these points without modification of the machine itself, by pure paper
work, eventually resulting in feeding in appropriate instructions.

To return to the various parts of the machine. I was saying that it will
work in the scale of two. It is not unnatural to use the convention that an
electrical pulse shall represent the digit 1 and that absence of a pulse shall
represent a digit 0. Thus a sequence of digits 0010110 would be represented
by a signal like

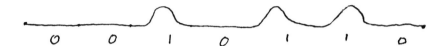

where the time interval might be one microsecond. Let us now look at what the process of binary addition is like. In ordinary decimal addition we always begin from the right, and the same naturally applies to binary. We have to do this because we cannot tell whether to carry unless we have already dealt with the less significant columns. The same applies with electronic addition, and therefore it is convenient to use the convention that if a sequence of pulses is coming down a line, then the least significant pulse always comes first. This has the unfortunate result that we must either write the least significant digit on the left in our binary numbers or else make time flow from right to left in our diagrams. As the latter alternative would involve writing from right to left as well as adding in that way, we have decided to put the least significant digit on the left. Now let us do a typical addition. Let us write the carry digits above the addends.

```
Carry     0  1  1  1  1  1  0  0  1  1
    A  0  1  1  0  1  1  0  0  1  0  1...
    B  0  1  1  1  0  1  0  0  1  1  ...
       ─────────────────────────────
          0  1  0  0  1  1  0  0  0
```

Note that I can do the addition only looking at a small part of the data. To do the addition electronically we need to produce a circuit with three inputs and two outputs.

Inputs		*Outputs*	
Addend A	α	Sum	δ
Addend B	β	Carry	ε
Carry from last column	γ		

This circuit must be such that

If no. of 1's on inputs α, β, γ is $\begin{cases} 0 & \text{Then sum} & 0 & \text{and} & 0 \\ 1 & \delta & 1 & \text{carry} & 0 \\ 2 & \text{is} & 0 & \varepsilon & 1 \\ 3 & & 1 & \text{is} & 1 \end{cases}$

It is very easy to produce a voltage proportional to the number of pulses on

the inputs, and one then merely has to provide a circuit which will discrimi-
nate between four different levels and put out the appropriate sum and
carry digits. I will not attempt to describe such a circuit; it can be quite
simple. When we are given the circuit we merely have to connect it up with
feedback and it is an adder. Thus:

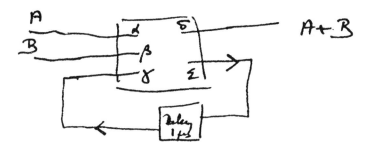

It will be seen that we have made use of the fact that the same process
is used in addition with each digit, and also the fact that the properties of
the electrical circuit are invariant under time shifts, at any rate if these are
multiples of the clock period. It might be said that we have made use of
the isomorphism between the group of these time shifts and the multiplica-
tive group of real numbers to simplify our apparatus, though I doubt if
many other applications of this principle could be found.

It will be seen that with such an adder the addition is broken down into
the most elementary steps possible, such as adding one and one. Each of
these occupies a microsecond. Our numbers will normally consist of 32
binary digits, so that two of them can be added in 32 microseconds.
Likewise we shall do multiplications in the form of a number of consecutive
additions of one and one or one and zero etc. There are 1024 such additions
or thereabouts to be done in a multiplication of one 32 digit number by
another, so that one might expect a multiplication to take about a millisecond. Actually the multiplier to be used on ACE will take rather over two
milliseconds. This may sound rather long, when the unit operation is only a
microsecond, but it actually seems that the machine is fairly well balanced
in this respect, i.e. the multiplication time is not a serious bottleneck.
Computers always spend just as long in writing numbers down and deciding what to do next as they do in actual multiplications, and it is just the
same with the ACE. A great deal of time is spent in getting numbers in and

out of storage and deciding what to do next. To complete the four elementary processes, subtraction is done by complementation and addition, and division is done by the use of the iteration formula

$$u_n = u_{n-1} + u_{n-1}(1 - au_{n-1})$$

[[12]] u_n converges to a^{-1} provided $|1 - au_0| < 1$. The error is squared at each step, so that the convergence is very rapid. This process is of course programmed, i.e. the only extra apparatus required is the delay lines required for storing the relevant instructions.

Passing on from the arithmetic part there remains the input and output. For this purpose we have chosen Hollerith card equipment. We are able to obtain this without having to do any special development work. The speeds obtainable are not very impressive compared with the speeds at which the electronic equipment works, but they are quite sufficient in all cases where the calculation is long and the result concise: the interesting cases in fact. It might appear that there would be a difficulty in converting the information provided at the slow speeds appropriate to the Hollerith equipment to the high speeds required with the ACE, but it is really quite easy. The Hollerith speeds are so slow as to be counted zero or stop for many purposes, and the problem reduces to the simple one of converting a number of statically given digits into a stream of pulses. This can be done by means of a form of electronic commutator.

Before leaving the outline of the description of the machine I should mention some of the tactical situations that are met with in programming. I can illustrate two of them in connection with the calculation of the reciprocal described above. One of these is the idea of the iterative cycle. Each time that we go from u_r to u_{r+1} we apply the same sequence of operations, and it will therefore be economical in storage space if we use the same instructions. Thus we go round and round a cycle of instructions:

It looks however as if we were in danger of getting stuck in this cycle, and unable to get out. The solution of this difficulty involves another tactical idea, that of 'discrimination' i.e. of deciding what to do next partly according to the results of the machine itself, instead of according to data available to the programmer. In this case we include a discrimination in each cycle, which takes us out of the cycle when the value of $|1 - au|$ is sufficiently small. It is like an aeroplane circling over an aerodrome, and asking permission to land after each circle. This is a very simple idea, but is of the utmost importance. The idea of the iterative cycle of instructions will also be seen to be rather fundamental when it is realised that the majority of the instructions in the memory must be obeyed a great number of times. If the whole memory were occupied by instructions, none of it being used for numbers or other data, and if each instruction were obeyed once only, but took the longest possible time, the machine could only remain working for sixteen seconds.

Another important idea is that of constructing an instruction and then obeying it. This can be used amongst other things for discrimination. In the example I have just taken for instance we could calculate a quantity which was 1 if $|1 - au|$ was less than $2^{-3.1}$ and 0 otherwise. By adding this quantity to the instruction that is obeyed at the forking point the instruction can be completely altered in its effect when finally $1 - au$ is reduced to sufficiently small dimensions.

Probably the most important idea involved in instruction tables is that of standard *subsidiary tables*. Certain processes are used repeatedly in all sorts of different connections, and we wish to use the same instructions, from the same part of the memory every time. Thus we may use interpolation for the calculation of a great number of different functions, but we shall always use the same instruction table for interpolation. We have only to think out how this is to be done once, and forget then how it is done. Each time we want to do an interpolation we have only to remember the memory position where this table is kept, and make the appropriate reference in the instruction table which is using the interpolation. We might for instance be making up an instruction table for finding values of $J_0(x)$ and use the interpolation table in this way. We should then say that the interpolation table was a subsidiary to the table for calculating $J_0(x)$. There is thus a sort of hierarchy ⟦13⟧ of tables. The interpolation table might be regarded as taking its orders from the J_0 table, and reporting its answers back to it. The master servant analogy is however not a very good one, as there are many more masters than servants, and many masters have to share the same servants.

Now let me give a picture of the operation of the machine. Let us begin with some problem which has been brought in by a customer. It will first go to the problems preparation section where it is examined to see whether it is in a suitable form and self-consistent, and a very rough computing procedure made out. It then goes to the tables preparation section. Let us suppose for example that the problem was to tabulate solutions of the equation

$$y'' + xy' = J_0(x)$$

with initial conditions $x = y = 0$, $y' = a$. This would be regarded as a particular case of solving the equation

$$y'' = F(x, y, y')$$

for which one would have instruction tables already prepared. One would need also a table to produce the function $F(x, y, z)$ (in this case $F(x, y, z) = J_0(x) - xz$ which would mainly involve a table to produce $J_0(x)$, and this we might expect to get off the shelf). A few additional details about the boundary conditions and the length of the arc would have to be dealt with, but much of this detail would also be found on the shelf, just like the table for obtaining $J_0(x)$. The instructions for the job would therefore consist of a considerable number taken off the shelf together with a few made up specially for the job in question. The instruction cards for the standard processes would have already been punched, but the new ones would have to be done separately. When these had all been assembled and checked they would be taken to the input mechanism, which is simply a Hollerith card feed. They would be put into the card hopper and a button pressed to start the cards moving through. It must be remembered that initially there are no instructions in the machine, and one's normal facilities are therefore not [[14]] available. The first few cards that pass in have therefore to be carefully thought out to deal with this situation. They are the initial input cards and are always the same. When they have passed in a few rather fundamental instruction tables will have been set up in the machine, including sufficient to enable the machine to read the special pack of cards that has been prepared for the job we are doing. When this has been done there are various possibilities as to what happens next, depending on the way the job has been programmed. The machine might have been made to go straight on through, and carry out the job, punching or printing all the answers required, and stopping when all of this has been done. But more probably it will have been arranged that the machine stops as soon as the instruction tables have been put in. This allows for the possibility of checking that the content of the memories is correct, and for a number of variations of procedure. It is clearly a suitable moment for a break. We might also make a number of other breaks. For instance we might be interested in certain particular values of the parameter a, which were experimentally obtained figures, and it would then be convenient to pause after each parameter value, and feed the next parameter value in from another card. Or one might prefer to have the cards all ready in the hopper and let the ACE take them in as it wanted them. One can do as one wishes, but one must make up one's mind. Each time the machine pauses in this way a 'word' or sequence of 32 binary digits is displayed on neon bulbs. This word indicates the reason for stopping. I have already mentioned two possible reasons. A large class of further possible reasons is provided by the checks. The programming should be done in such a way that the ACE is frequently investigating identities which should be satisfied if all is as it should be.   [[15]] Whenever one of these checks fails the machine stops and displays a word which describes what check has failed.

It will be seen that the possibilities as to what one may do are immense. One of our difficulties will be the maintainence of an appropriate discipline, so that we do not lose track of what we are doing. We shall need a number of efficient librarian types to keep us in order.

Finally I should like to make a few conjectures as to the repercussions that electronic digital computing machinery will have on mathematics. I have already mentioned that the ACE will do the work of about 10,000 computers. It is to be expected therefore that large scale hand-computing will die out. Computers will still be employed on small calculations, such as the substitution of values in formulae, but whenever a single calculation may be expected to take a human computer days of work, it will presumably be done by an electronic computer instead. This will not necessitate everyone interested in such work having an electronic computer. It would be quite possible to arrange to control a distant computer by means of a telephone line. Special input and output machinery would be developed for use at these out stations, and would cost a few hundred pounds at most. The main bulk of the work done by these computers will however consist of problems which could not have been tackled by hand computing because of the scale of the undertaking. In order to supply the machine with these problems we shall need a great number of mathematicians of ability. These mathematicians will be needed in order to do the preliminary research on the problems, putting them into a form for computation. There will be considerable scope for analysts. When a human computer is working on a problem he can usually apply some common sense to give him an idea of how accurate his answers are. With a digital computer we can no longer rely on common sense, and the bounds of error must be based on some proved inequalities. We need analysts to find the appropriate inequalities for us. The inequalities need not always be explicit, i.e. one need not have them in such a form that we can tell, before the calculation starts, and using only pencil and paper, how big the error will be. The error calculation may be a serious part of the ACE's duties. To an extent it may be possible to replace the estimates of error by statistical estimates obtained by repeating the job several times, and doing the rounding off differently each time, controlling it by some random element, some electronic roulette wheel. Such statistical estimates however leave much in doubt, are wasteful in machine time, and give no indication of what can be done if it turns out that the errors are intolerably large. The statistical method can only help the analyst, not replace him.

[16]

Analysis is just one of the purposes for which we shall need good mathematicians. Roughly speaking those who work in connection with the ACE will be divided into its masters and its servants. Its masters will plan out instruction tables for it, thinking up deeper and deeper ways of using it. Its servants will feed it with cards as it calls for them. They will put right any parts that go wrong. They will assemble data that it requires. In fact the servants will take the place of limbs. As time goes on the calculator itself will take over the functions both of masters and of servants. The servants will be replaced by mechanical and electrical limbs and sense organs. One might for instance provide curve followers to enable data to be taken direct from curves instead of having girls read off values and punch them on cards. The masters are liable to get replaced because as soon as any technique becomes at all stereotyped it becomes possible to devise a system of instruction tables which will enable the electronic computer to do it for itself. It may happen however that the masters will refuse to do this. They may be unwilling to let their jobs be stolen from them in this way. In that case they would surround the whole of their work with mystery and make excuses, couched in well chosen gibberish, whenever any dangerous suggestions were made. I think that a reaction of this kind is a very real danger. This topic naturally leads to the question as to how far it is possible in

[102]

principle for a computing machine to simulate human activities. I will return to this later, when I have discussed the effects of these machines on mathematics a little further.

I expect that digital computing machines will eventually stimulate a considerable interest in symbolic logic and mathematical philosophy. The language in which one communicates with these machines, i.e. the language of instruction tables, forms a sort of symbolic logic. The machine interprets whatever it is told in a quite definite manner without any sense of humour or sense of proportion. Unless in communicating with it one says exactly what one means, trouble is bound to result. Actually one could communicate with these machines in any language provided it was an exact language, i.e. in principle one should be able to communicate in any symbolic logic, provided that the machine were given instruction tables which would enable it to interpret that logical system. This would mean that there will be much more practical scope for logical systems than there has been in the past. Some attempts will probably be made to get the machine to do actual [[17]] manipulations of mathematical formulae. To do so will require the development of a special logical system for the purpose. This system should resemble normal mathematical procedure closely, but at the same time should be as unambiguous as possible. As regards mathematical philosophy, since the machines will be doing more and more mathematics themselves, the centre of gravity of the human interest will be driven further and further into philosophical questions of what can in principle be done etc.

It has been said that computing machines can only carry out the processes that they are instructed to do. This is certainly true in the sense that if they do something other than what they were instructed then they have just made some mistake. It is also true that the intention in constructing these machines in the first instance is to treat them as slaves, giving them only jobs which have been thought out in detail, jobs such that the user of the machine fully understands what in principle is going on all the time. Up till the present machines have only been used in this way. But is it necessary that they should always be used in such a manner? Let us suppose we have set up a machine with certain initial instruction tables, so constructed that these tables might on occasion, if good reason arose, modify those tables. [[18]] One can imagine that after the machine had been operating for some time, the instructions would have altered out of all recognition, but nevertheless still be such that one would have to admit that the machine was still doing very worthwhile calculations. Possibly it might still be getting results of the type desired when the machine was first set up, but in a much more efficient manner. In such a case one would have to admit that the progress of the machine had not been foreseen when its original instructions were put in. It would be like a pupil who had learnt much from his master, but had added much more by his own work. When this happens I feel that one is obliged to regard the machine as showing intelligence. As soon as one can provide a reasonably large memory capacity it should be possible to begin to experiment on these lines. The memory capacity of the human brain is probably of the order of ten thousand million binary digits. But most of this is probably used in remembering visual impressions, and other comparatively wasteful ways. One might reasonably hope to be able to make some real progress with a few million digits, especially if one confined one's investigations to some rather limited field such as the game of chess. It would probably be quite easy to find instruction tables which would enable the ACE to win against an average player. Indeed Shannon of Bell Telephone laboratories tells me that he has won games playing by rule of thumb: the skill of his opponents is not stated. But I would not consider [[19]] such a victory very significant. What we want is a machine that can learn from experience. The possibility of letting the machine alter its own instructions provides the mechanism for this, but this of course does not get us very far.

It might be argued that there is a fundamental contradiction in the idea of a machine with intelligence. It is certainly true that 'acting like a machine', has become synonymous with lack of adaptability. But the reason for this is obvious. Machines in the past have had very little storage, and there has been no question of the machine having any discretion. The argument might however be put into a more aggressive form. It has for instance been shown that with certain logical systems there can be no machine which will distinguish provable formulae of the system from unprovable, i.e. that there is no test that the machine can apply which will divide propositions with certainty into these two classes. Thus if a machine is made for this purpose it must in some cases fail to give an answer. On the other hand if a mathematician is confronted with such a problem he would search around and find new methods of proof, so that he ought eventually to be able to reach a decision about any given formula. This would be the argument. Against it I would say that fair play must be given to the machine. Instead of it sometimes giving no answer we could arrange that it

[[104]]

gives occasional wrong answers. But the human mathematician would likewise make blunders when trying out new techniques. It is easy for us to regard these blunders as not counting and give him another chance, but the machine would probably be allowed no mercy. In other words then, if a machine is expected to be infallible, it cannot also be intelligent. There are several mathematical theorems which say almost exactly that. But these theorems say nothing about how much intelligence may be displayed if a machine makes no pretence at infallibility. To continue my plea for 'fair play for the machines' when testing their I.Q. A human mathematician has always undergone an extensive training. This training may be regarded as not unlike putting instruction tables into a machine. One must therefore not expect a machine to do a very great deal of building up of instruction tables on its own. No man adds very much to the body of knowledge, why should we expect more of a machine? Putting the same point differently, the machine must be allowed to have contact with human beings in order that it may adapt itself to their standards. The game of chess may perhaps be rather suitable for this purpose, as the moves of the machine's opponent will automatically provide this contact.

Intelligent Machinery

A. M. Turing
[1912—1954]

Abstract

The possible ways in which machinery might be made to show intelligent behaviour are discussed. The analogy with the human brain is used as a guiding principle. It is pointed out that the potentialities of the human intelligence can only be realized if suitable education is provided. The investigation mainly centres round an analogous teaching process applied to machines. The idea of an unorganized machine is defined, and it is suggested that the infant human cortex is of this nature. Simple examples of such machines are given, and their education by means of rewards and punishments is discussed. In one case the education process is carried through until the organization is similar to that of an ACE.

I propose to investigate the question as to whether it is possible for machinery to show intelligent behaviour. It is usually assumed without argument that it is not possible. Common catch phrases such as 'acting like a machine', 'purely mechanical behaviour' reveal this common attitude. It is not difficult to see why such an attitude should have arisen. Some of the reasons are:

(a) An unwillingness to admit the possibility that mankind can have any rivals in intellectual power. This occurs as much amongst intellectual people as amongst others: they have more to lose. Those who admit the possibility all agree that its realization would be very disagreeable. The same situation arises in connection with the possibility of our being superseded by some other animal species. This is almost as disagreeable and its theoretical possibility is indisputable.

(b) A religious belief that any attempt to construct such machines is a sort of Promethean irreverence.

(c) The very limited character of the machinery which has been used until recent times (e.g. up to 1940). This encouraged the belief that machinery was necessarily limited to extremely straightforward, possibly even to repetitive, jobs. This attitude is very well expressed by Dorothy Sayers (*The Mind of the Maker* p. 46) '... which imagines that God, having created his Universe, has now screwed the cap on His pen, put His feet on the

3

mantelpiece and left the work to get on with itself.' This, however, rather comes into St Augustine's category of figures of speech or enigmatic sayings framed from things which do not exist at all. We simply do not know of any creation which goes on creating itself in variety when the creator has withdrawn from it. The idea is that God simply created a vast machine and has left it working until it runs down from lack of fuel. This is another of those obscure analogies, since we have no experience of machines that produce variety of their own accord; the nature of a machine is to 'do the same thing over and over again so long as it keeps going'.

(d) Recently the theorem of Gödel and related results (Gödel 1931, Church 1936, Turing 1937) have shown that if one tries to use machines for such purposes as determining the truth or falsity of mathematical theorems and one is not willing to tolerate an occasional wrong result, then any given machine will in some cases be unable to give an answer at all. On the other hand the human intelligence seems to be able to find methods of ever-increasing power for dealing with such problems 'transcending' the methods available to machines.

[[1]] (e) In so far as a machine can show intelligence this is to be regarded as nothing but a reflection of the intelligence of its creator.

REFUTATION OF SOME OBJECTIONS

In this section I propose to outline reasons why we do not need to be influenced by the above-described objections. The objections (a) and (b), being purely emotional, do not really need to be refuted. If one feels it necessary to refute them there is little to be said that could hope to prevail, though the actual production of the machines would probably have some effect. In so far then as we are influenced by such arguments we are bound to be left feeling rather uneasy about the whole project, at any rate for the present. These arguments cannot be wholly ignored, because the idea of 'intelligence' is itself emotional rather than mathematical.

The objection (c) in its crudest form is refuted at once by the actual existence of machinery (ENIAC etc.) which can go on through immense numbers (e.g. $10^{60,000}$ about for ACE) of operations without repetition, assuming no breakdown. The more subtle forms of this objection will be considered at length on pages 18–22.

The argument from Gödel's and other theorems (objection d) rests essentially on the condition that the machine must not make mistakes. But this is not a requirement for intelligence. It is related that the infant Gauss was asked at school to do the addition $15+18+21+\ldots+54$ (or something of the kind) and that he immediately wrote down 483, presumably having

[[2]] calculated it as $(15+54)(54-12)/2.3$. One can imagine circumstances where a foolish master told the child that he ought instead to have added 18 to 15 obtaining 33, then added 21, etc. From some points of view this would be a 'mistake', in spite of the obvious intelligence involved. One can also

4

imagine a situation where the children were given a number of additions to do, of which the first 5 were all arithmetic progressions, but the 6th was say $23+34+45+\ldots+100+112+122+\ldots+199$. Gauss might have given the answer to this as if it were an arithmetic progression, not having noticed that the 9th term was 112 instead of 111. This would be a definite mistake, which the less intelligent children would not have been likely to make.

The view (d) that intelligence in machinery is merely a reflection of that of its creator is rather similar to the view that the credit for the discoveries of a pupil should be given to his teacher. In such a case the teacher would be pleased with the success of his methods of education, but would not claim the results themselves unless he had actually communicated them to his pupil. He would certainly have envisaged in very broad outline the sort of thing his pupil might be expected to do, but would not expect to foresee any sort of detail. It is already possible to produce machines where this sort of situation arises in a small degree. One can produce 'paper machines' for playing chess. Playing against such a machine gives a definite feeling that one is pitting one's wits against something alive.   ⟦3⟧

These views will all be developed more completely below.

VARIETIES OF MACHINERY

It will not be possible to discuss possible means of producing intelligent machinery without introducing a number of technical terms to describe different kinds of existent machinery.

'Discrete' and *'continuous'* *machinery.* We may call a machine 'discrete' when it is natural to describe its possible states as a discrete set, the motion of the machine occurring by jumping from one state to another. The states of 'continuous' machinery on the other hand form a continuous manifold, and the behaviour of the machine is described by a curve on this manifold. All machinery can be regarded as continuous, but when it is possible to regard it as discrete it is usually best to do so. The states of discrete machinery will be described as 'configurations'.

'Controlling' and *'active'* *machinery.* Machinery may be described as 'controlling' if it only deals with information. In practice this condition is much the same as saying that the magnitude of the machine's effects may be as small as we please, so long as we do not introduce confusion through Brownian movement, etc. 'Active' machinery is intended to produce some definite physical effect.

Examples	A Bulldozer	Continuous Active	
	A Telephone	Continuous Controlling	
	A Brunsviga	Discrete Controlling	⟦4⟧
	A Brain (probably)	Continuous Controlling, but is very similar to much discrete machinery	

5

The ENIAC, ACE, etc.   Discrete Controlling
A Differential Analyser   Continuous Controlling.

We shall mainly be concerned with discrete controlling machinery. As we have mentioned, brains very nearly fall into this class, and there seems every reason to believe that they could have been made to fall genuinely into it without any change in their essential properties. However, the property of being 'discrete' is only an advantage for the theoretical investigator, and serves no evolutionary purpose, so we could not expect Nature to assist us by producing truly 'discrete' brains.

Given any discrete machine the first thing we wish to find out about it is the number of states (configurations) it can have. This number may be infinite (but enumerable) in which case we say that the machine has infinite memory (or storage) capacity. If the machine has a finite number N of possible states then we say that it has a memory capacity of (or equivalent to) $\log_2 N$ binary digits. According to this definition we have the following table of capacities, very roughly

Brunsviga	90
ENIAC without cards and with fixed programme	600
ACE as proposed	60,000
Manchester machine (as actually working 8 August 1947)	1,100

The memory capacity of a machine more than anything else determines the complexity of its possible behaviour.

The behaviour of a discrete machine is completely described when we are given the state (configuration) of the machine as a function of the immediately preceding state and the relevant external data.

Logical computing machines (LCMs)

In Turing (1937) a certain type of discrete machine was described. It had an infinite memory capacity obtained in the form of an infinite tape marked out into squares on each of which a symbol could be printed. At any moment there is one symbol in the machine; it is called the scanned symbol. The machine can alter the scanned symbol and its behaviour is in part described by that symbol, but the symbols on the tape elsewhere do not affect the behaviour of the machine. However the tape can be moved back and forth through the machine, this being one of the elementary operations of the machine. Any symbol on the tape may therefore eventually have an innings.

These machines will here be called 'Logical Computing Machines'. They are chiefly of interest when we wish to consider what a machine could in principle be designed to do, when we are willing to allow it both unlimited time and unlimited storage capacity.

Universal logical computing machines. It is possible to describe LCMs in a very standard way, and to put the description into a form which can be

6

'understood' (i.e., applied by) a special machine. In particular it is possible to design a 'universal machine' which is an LCM such that if the standard description of some other LCM is imposed on the otherwise blank tape from outside, and the (universal) machine then set going it will carry out the operations of the particular machine whose description it was given. For details the reader must refer to Turing (1937).

The importance of the universal machine is clear. We do not need to have an infinity of different machines doing different jobs. A single one will suffice. The engineering problem of producing various machines for various jobs is replaced by the office work of 'programming' the universal machine to do these jobs.

It is found in practice that LCMs can do anything that could be described as 'rule of thumb' or 'purely mechanical'. This is sufficiently well established   ⟦5⟧ that it is now agreed amongst logicians that 'calculable by means of an LCM' is the correct accurate rendering of such phrases. There are several mathematically equivalent but superficially very different renderings.

Practical computing machines (PCMs)

Although the operations which can be performed by LCMs include every rule-of-thumb process, the number of steps involved tends to be enormous. This is mainly due to the arrangement of the memory along the tape. Two facts which need to be used together may be stored very far apart on the tape. There is also rather little encouragement, when dealing with these machines, to condense the stored expressions at all. For instance the number of symbols required in order to express a number in Arabic form (e.g., 149056) cannot be given any definite bound, any more than if the numbers are expressed in the 'simplified Roman' form (IIIII . . . I, with 149056 occurrences of I). As the simplified Roman system obeys very much simpler laws one uses it instead of the Arabic system.

In practice however one *can* assign finite bounds to the numbers that one will deal with. For instance we can assign a bound to the number of steps that we will admit in a calculation performed with a real machine in the following sort of way. Suppose that the storage system depends on charging condensers of capacity $C = 1\,\mu f$, and that we use two states of charging, $E = 100$ volts and $-E = -100$ volts. When we wish to use the information carried by the condenser we have to observe its voltage. Owing to thermal agitation the voltage observed will always be slightly wrong, and the probability of an error between V and $V - dV$ volts is

$$\frac{2kT}{\pi C}\, e^{-\frac{1}{2}V^2 C/kT}\, V\mathrm{d}V$$

where k is Boltzmann's constant. Taking the values suggested we find that the probability of reading the sign of the voltage wrong is about $10^{-1\cdot2\times10^{16}}$. If then a job took more than $10^{10^{17}}$ steps we should be virtually certain of

7

getting the wrong answer, and we may therefore restrict ourselves to jobs with fewer steps. Even a bound of this order might have useful simplifying effects. More practical bounds are obtained by assuming that a light wave must travel at least 1 cm between steps (this would only be false with a very small machine), and that we could not wait more than 100 years for an answer. This would give a limit of 10^{20} steps. The storage capacity will probably have a rather similar bound, so that we could use sequences of 20 decimal digits for describing the position in which a given piece of data was to be found, and this would be a really valuable possibility.

Machines of the type generally known as 'Automatic Digital Computing Machines' often make great use of this possibility. They also usually put a great deal of their stored information in a form very different from the tape form. By means of a system rather reminiscent of a telephone exchange it is made possible to obtain a piece of information almost immediately by 'dialling' the position of this information in the store. The delay may be only a few microseconds with some systems. Such machines will be described as 'Practical Computing Machines'.

Universal practical computing machines. Nearly all of the PCMs now under construction have the essential properties of the 'Universal Logical Computing Machines' mentioned earlier. In practice, given any job which could have been done on an LCM one can also do it on one of these digital computers. I do not mean that we can do any required job of the type mentioned on it by suitable programming. The programming is pure paper work. It naturally occurs to one to ask whether, e.g., the ACE would be truly universal if its memory capacity were infinitely extended. I have investigated this question, and the answer appears to be as follows, though I have not proved any formal mathematical theorem about it. As has been explained, the ACE at present uses finite sequences of digits to describe positions in its memory: they are actually sequences of 9 binary digits (September 1947). The ACE also works largely for other purposes with sequences of 32 binary digits. If the memory were extended, e.g., to 1000 times its present capacity, it would be natural to arrange the memory in blocks of nearly the maximum capacity which can be handled with the 9 digits, and from time to time to switch from block to block. A relatively small part would never be switched. This would contain some of the more fundamental instruction tables and those concerned with switching. This part might be called the 'central part'. One would then need to have a number which described which block was in action at any moment. However this number might be as large as one pleased. Eventually the point would be reached where it could not be stored in a word (32 digits), or even in the central part. One would then have to set aside a block for storing the number, or even a sequence of blocks, say blocks 1, 2, ... n. We should then have to store n, and in theory it would be of indefinite size. This sort of process can be extended in all sorts of ways, but we shall always be left with a positive integer which is of indefinite size

8

⟦6⟧

and which needs to be stored somewhere, and there seems to be no way out of the difficulty but to introduce a 'tape'. But once this has been done, and since we are only trying to prove a theoretical result, one might as well, whilst proving the theorem, ignore all the other forms of storage. One will in fact have a ULCM with some complications. This in effect means that one will not be able to prove any result of the required kind which gives any intellectual satisfaction.

Paper machines

It is possible to produce the effect of a computing machine by writing down a set of rules of procedure and asking a man to carry them out. Such a combination of a man with written instructions will be called a 'Paper Machine'. A man provided with paper, pencil, and rubber, and subject to strict discipline, is in effect a universal machine. The expression 'paper machine' will often be used below.

Partially random and apparently partially random machines

It is possible to modify the above described types of discrete machines by allowing several alternative operations to be applied at some points, the alternatives to be chosen by a random process. Such a machine will be described as 'partially random'. If we wish to say definitely that a machine is not of this kind we will describe it as 'determined'. Sometimes a machine may be strictly speaking determined but appear superficially as if it were partially random. This would occur if for instance the digits of the number π were used to determine the choices of a partially random machine, where previously a dice thrower or electronic equivalent had been used. These machines are known as apparently partially random.

UNORGANIZED MACHINES

So far we have been considering machines which are designed for a definite purpose (though the universal machines are in a sense an exception). We might instead consider what happens when we make up a machine in a comparatively unsystematic way from some kind of standard components. We could consider some particular machine of this nature and find out what sort of things it is likely to do. Machines which are largely random in their construction in this way will be called 'Unorganized Machines'. This does not pretend to be an accurate term. It is conceivable that the same machine might be regarded by one man as organized and by another as unorganized.

A typical example of an unorganized machine would be as follows. The machine is made up from a rather large number N of similar units. Each unit has two input terminals, and has an output terminal which can be connected to the input terminals of (0 or more) other units. We may imagine that for each integer r, $1 \leqslant r \leqslant N$ two numbers $i(r)$ and $j(r)$ are chosen at random

9

from 1 ... N and that we connect the inputs of unit r to the outputs of units (r) and $j(r)$. All of the units are connected to a central synchronizing unit from which synchronizing pulses are emitted at more or less equal intervals of time. The times when these pulses arrive will be called 'moments'. Each unit is capable of having two states at each moment. These states may be called 0 and 1. The state is determined by the rule that the states of the units from which the input leads come are to be taken at the previous moment, multiplied together and the result subtracted from 1. An unorganized machine of this character is shown in the diagram below.

r	i(r)	j(r)
1	3	2
2	3	5
3	4	5
4	3	4
5	2	5

A sequence of six possible consecutive conditions for the whole machine is:

[[7]]

1	1	1	0	0	1	0
2	1	1	1	0	1	0
3	0	1	1	1	1	1
4	0	1	0	1	0	1
5	1	0	1	0	1	0

The behaviour of a machine with so few units is naturally very trivial. However, machines of this character can behave in a very complicated manner when the number of units is large. We may call these A-type unorganized machines. Thus the machine in the diagram is an A-type unorganized machine of 5 units. The motion of an A-type machine with N units is of course eventually periodic, as in any determined machine with finite memory capacity. The period cannot exceed 2^N moments, nor can the length of time before the periodic motion begins. In the example above the period is 2 moments and there are 3 moments before the periodic motion begins. 2^N is 32.

The A-type unorganized machines are of interest as being about the simplest model of a nervous system with a random arrangement of neurons. It would therefore be of very great interest to find out something about their behaviour. A second type of unorganized machine will now be described, not because it is

10

of any great intrinsic importance, but because it will be useful later for illustrative purposes. Let us denote the circuit

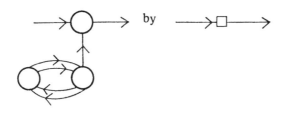

as an abbreviation. Then for each A-type unorganized machine we can construct another machine by replacing each connection ———>——— in it by ———>☐———>. The resulting machines will be called B-type unorganized machines. It may be said that the B-type machines are all A-type. To this I would reply that the above definitions if correctly (but drily!) set out would take the form of describing the probability of an A- (or B-) type machine belonging to a given set; it is not merely a definition of which are the A-type machines and which are the B-type machines. If one chooses an A-type machine, with a given number of units, at random, it will be extremely unlikely that one will get a B-type machine.

It is easily seen that the connection ———>☐———> can have three conditions. It may (i) pass all signals through with interchange of 0 and 1, or (ii) it may convert all signals into 1, or again (iii) it may act as in (i) and (ii) in alternate moments. (Alternative (iii) has two sub-cases.) Which of these cases applies depends on the initial conditions. There is a delay of two moments in going through ———>☐———> .

INTERFERENCE WITH MACHINERY. MODIFIABLE AND SELF-MODIFYING MACHINERY

The types of machine that we have considered so far are mainly ones that are allowed to continue in their own way for indefinite periods without interference from outside. The universal machines were an exception to this, in that from time to time one might change the description of the machine which is being imitated. We shall now consider machines in which such interference is the rule rather than the exception.

We may distinguish two kinds of interference. There is the extreme form in which parts of the machine are removed and replaced by others. This may be described as 'screwdriver interference'. At the other end of the scale is 'paper interference', which consists in the mere communication of information to the machine, which alters its behaviour. In view of the properties of the universal machine we do not need to consider the difference between these

11

two kinds of machine as being so very radical after all. Paper interference when applied to the universal machine can be as useful as screwdriver interference.

We shall mainly be interested in paper interference. Since screwdriver interference can produce a completely new machine without difficulty there is rather little to be said about it. In future 'interference' will normally mean 'paper interference'.

When it is possible to alter the behaviour of a machine very radically we may speak of the machine as being 'modifiable'. This is a relative term. One machine may be spoken of as being more modifiable than another.

One may also sometimes speak of a machine modifying itself, or of a machine changing its own instructions. This is really a nonsensical form of phraseology, but is convenient. Of course, according to our conventions the 'machine' is completely described by the relation between its possible configurations at consecutive moments. It is an abstraction which, by the form of its definition, cannot change in time. If we consider the machine as starting in a particular configuration, however, we may be tempted to ignore those configurations which cannot be reached without interference from it. If we do this we should get a 'successor relation' for the configurations with different properties from the original one and so a different 'machine'.

If we now consider interference, we should say that each time interference occurs the machine is probably changed. It is in this sense that interference 'modifies' a machine. The sense in which a machine can modify itself is even more remote. We may, if we wish, divide the operations of the machine into two classes, normal and self-modifying operations. So long as only normal operations are performed we regard the machine as unaltered. Clearly the idea of 'self-modification' will not be of much interest except where the division of operations into the two classes is made very carefully. The sort of case I have in mind is a computing machine like the ACE where large parts of the storage are normally occupied in holding instruction tables. (Instruction tables are the equivalent in UPCMs of descriptions of machines in ULCMs). Whenever the content of this storage was altered by the internal operations of the machine, one would naturally speak of the machine 'modifying itself'.

MAN AS A MACHINE

A great positive reason for believing in the possibility of making thinking machinery is the fact that it is possible to make machinery to imitate any small part of a man. That the microphone does this for the ear, and the television camera for the eye are commonplaces. One can also produce remote-controlled robots whose limbs balance the body with the aid of servo-mechanisms. Here we are chiefly interested in the nervous system. We could produce fairly accurate electrical models to copy the behaviour of nerves, but there seems very little point in doing so. It would be rather like

12

putting a lot of work into cars which walked on legs instead of continuing to use wheels. The electrical circuits which are used in electronic computing machinery seem to have the essential properties of nerves. They are able to transmit information from place to place, and also to store it. Certainly the nerve has many advantages. It is extremely compact, does not wear out (probably for hundreds of years if kept in a suitable medium!) and has a very low energy consumption. Against these advantages the electronic circuits have only one counter-attraction, that of speed. This advantage is, however, on such a scale that it may possibly outweigh the advantages of the nerve.

One way of setting about our task of building a 'thinking machine' would be to take a man as a whole and to try to replace all the parts of him by machinery. He would include television cameras, microphones, loudspeakers, wheels and 'handling servo-mechanisms' as well as some sort of 'electronic brain'. This would be a tremendous undertaking of course. The object, if produced by present techniques, would be of immense size, even if the 'brain' part were stationary and controlled the body from a distance. In order that the machine should have a chance of finding things out for itself it should be allowed to roam the countryside, and the danger to the ordinary citizen would be serious. Moreover even when the facilities mentioned above were provided, the creature would still have no contact with food, sex, sport and many other things of interest to the human being. Thus although this method is probably the 'sure' way of producing a thinking machine it seems to be altogether too slow and impracticable.

Instead we propose to try and see what can be done with a 'brain' which is more or less without a body providing, at most, organs of sight, speech, and hearing. We are then faced with the problem of finding suitable branches of thought for the machine to exercise its powers in. The following fields appear to me to have advantages:

(i) Various games, e.g., chess, noughts and crosses, bridge, poker

(ii) The learning of languages

(iii) Translation of languages

(iv) Cryptography

(v) Mathematics.

Of these (i), (iv), and to a lesser extent (iii) and (v) are good in that they require little contact with the outside world. For instance in order that the machine should be able to play chess its only organs need be 'eyes' capable of distinguishing the various positions on a specially made board, and means for announcing its own moves. Mathematics should preferably be restricted to branches where diagrams are not much used. Of the above possible fields the learning of languages would be the most impressive, since it is the most human of these activities. This field seems however to depend rather too much on sense organs and locomotion to be feasible.

13

The field of cryptography will perhaps be the most rewarding. There is a remarkably close parallel between the problems of the physicist and those of the cryptographer. The system on which a message is enciphered corresponds to the laws of the universe, the intercepted messages to the evidence available, the keys for a day or a message to important constants which have to be determined. The correspondence is very close, but the subject matter of cryptography is very easily dealt with by discrete machinery, physics not so easily.

EDUCATION OF MACHINERY

Although we have abandoned the plan to make a 'whole man', we should be wise to sometimes compare the circumstances of our machine with those of a man. It would be quite unfair to expect a machine straight from the factory to compete on equal terms with a university graduate. The graduate has had contact with human beings for twenty years or more. This contact has been modifying his behaviour pattern throughout that period. His teachers have been intentionally trying to modify it. At the end of the period a large number [[8]] of standard routines will have been superimposed on the original pattern of his brain. These routines will be known to the community as a whole. He is then in a position to try out new combinations of these routines, to make slight variations on them, and to apply them in new ways.

We may say then that in so far as a man is a machine he is one that is subject to very much interference. In fact interference will be the rule rather than the exception. He is in frequent communication with other men, and is continually receiving visual and other stimuli which themselves constitute a form of interference. It will only be when the man is 'concentrating' with a view to eliminating these stimuli or 'distractions' that he approximates a machine without interference.

We are chiefly interested in machines with comparatively little interference, for reasons given in the last section, but it is important to remember that although a man when concentrating may behave like a machine without interference, his behaviour when concentrating is largely determined by the way he has been conditioned by previous interference.

If we are trying to produce an intelligent machine, and are following the human model as closely as we can, we should begin with a machine with very little capacity to carry out elaborate operations or to react in a disciplined manner to orders (taking the form of interference). Then by applying appropriate interference, mimicking education, we should hope to modify the machine until it could be relied on to produce definite reactions to certain commands. This would be the beginning of the process. I will not attempt to follow it further now.

ORGANIZING UNORGANIZED MACHINERY

Many unorganized machines have configurations such that if once that configuration is reached, and if the interference thereafter is appropriately

14

restricted, the machine behaves as one organized for some definite purpose. For instance, the B-type machine shown below was chosen at random.

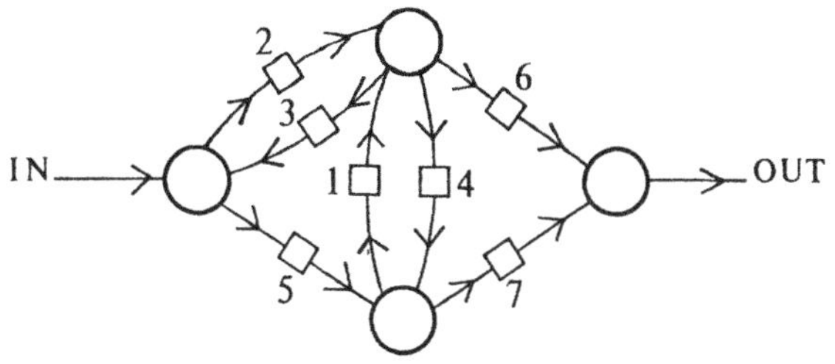

If the connections numbered 1, 3, 6, 4, are in condition (ii) initially and connections 2, 5, 7 are in condition (i), then the machine may be considered to be one for the purpose of passing on signals with a delay of 4 moments. This is a particular case of a very general property of B-type machines (and many other types), viz., that with suitable initial conditions they will do any required job, given sufficient time and provided the number of units is sufficient. In particular with a B-type unorganized machine with sufficient units one can find initial conditions which will make it into a universal machine with a given storage capacity. (A formal proof to this effect might be of some interest, or even a demonstration of it starting with a particular unorganized B-type machine, but I am not giving it as it lies rather too far outside the main argument.)

With these B-type machines the possibility of interference which could set in appropriate initial conditions has not been arranged for. It is however not difficult to think of appropriate methods by which this could be done. For instance instead of the connection

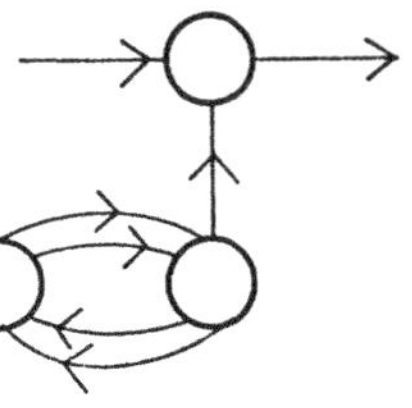

one might use

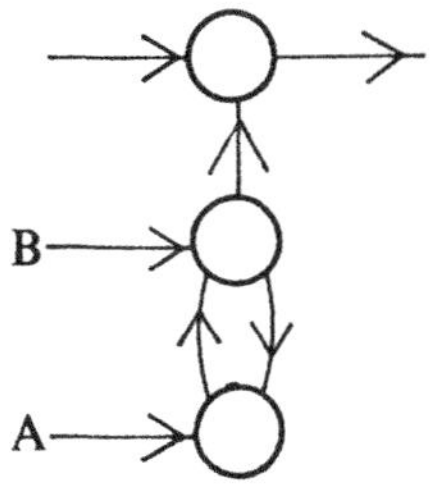

15

Here A, B are interfering inputs, normally giving the signal '1'. By supplying appropriate other signals at A, B we can get the connection into condition (i) or (ii), as desired. However this requires two special interfering inputs for each connection.

We shall be interested mainly in cases where there are only quite few independent inputs altogether, so that all the interference which sets up the 'initial conditions' of the machine has to be provided through one or two inputs. The process of setting up these initial conditions so that the machine will carry out some particular useful task may be called 'organizing the machine'. 'Organizing' is thus a form of 'modification'.

THE CORTEX AS AN UNORGANIZED MACHINE

Many parts of a man's brain are definite nerve circuits required for quite definite purposes. Examples of these are the 'centres' which control respiration, sneezing, following moving objects with the eyes, etc.: all the reflexes proper (not 'conditioned') are due to the activities of these definite structures in the brain. Likewise the apparatus for the more elementary analysis of shapes and sounds probably comes into this category. But the more intellectual activities of the brain are too varied to be managed on this basis. The difference between the languages spoken on the two sides of the Channel is not due to difference in development of the French-speaking and English-speaking parts of the brain. It is due to the linguistic parts having been subjected to different training. We believe then that there are large parts of the brain, chiefly in the cortex, whose function is largely indeterminate. In the infant these parts do not have much effect: the effect they have is unco-ordinated. In the adult they have great and purposive effect: the form of this effect depends on the training in childhood. A large remnant of the random behaviour of infancy remains in the adult.

All of this suggests that the cortex of the infant is an unorganized machine, which can be organized by suitable interfering training. The organizing might result in the modification of the machine into a universal machine or something like it. This would mean that the adult will obey orders given in appropriate language, even if they were very complicated; he would have no common sense, and would obey the most ridiculous orders unflinchingly. When all his orders had been fulfilled he would sink into a comatose state or perhaps obey some standing order, such as eating. Creatures not unlike this can really be found, but most people behave quite differently under many circumstance. However the resemblance to a universal machine is still very great, and suggests to us that the step from the unorganized infant to a universal machine is one which should be understood. When this has been mastered we shall be in a far better position to consider how the organizing process might have been modified to produce a more normal type of mind.

This picture of the cortex as an unorganized machine is very satisfactory

16

from the point of view of evolution and genetics. It clearly would not require any very complex system of genes to produce something like the A- or B-type unorganized machine. In fact this should be much easier than the production of such things as the respiratory centre. This might suggest that intelligent races could be produced comparatively easily. I think this is wrong because the possession of a human cortex (say) would be virtually useless if no attempt was made to organize it. Thus if a wolf by a mutation acquired a human cortex there is little reason to believe that he would have any selective advantage. If however the mutation occurred in a milieu where speech had developed (parrot-like wolves), and if the mutation by chance had well permeated a small community, then some selective advantage might be felt. It would then be possible to pass information on from generation to generation. However this is all rather speculative.

EXPERIMENTS IN ORGANIZING: PLEASURE–PAIN SYSTEMS

It is interesting to experiment with unorganized machines admitting definite types of interference and try to organize them, e.g., to modify them into universal machines.

The organization of a machine into a universal machine would be most impressive if the arrangements of interference involve very few inputs. The training of the human child depends largely on a system of rewards and punishments, and this suggests that it ought to be possible to carry through the organizing with only two interfering inputs, one for 'pleasure' or 'reward' (R) and the other for 'pain' or punishment' (P). One can devise a large number of such 'pleasure–pain' systems. I will use this term to mean an unorganized machine of the following general character: The configurations of the machine are described by two expressions, which we may call the character-expression and the situation-expression. The character and situation at any moment, together with the input signals, determine the character and situation at the next moment. The character may be subject to some random variation. Pleasure interference has a tendency to fix the character, i.e., towards preventing it changing, whereas pain stimuli tend to disrupt the character, causing features which had become fixed to change, or to become again subject to random variation.

This definition is probably too vague and general to be very helpful. The idea is that when the 'character' changes we like to think of it as a change in the machine, but the 'situation' is merely the configuration of the machine described by the character. It is intended that pain stimuli occur when the machine's behaviour is wrong, pleasure stimuli when it is particularly right. With appropriate stimuli on these lines, judiciously operated by the 'teacher', one may hope that the 'character' will converge towards the one desired, i.e., that wrong behaviour will tend to become rare.

I have investigated a particular type of pleasure–pain system, which I will now describe.

17

THE P-TYPE UNORGANIZED MACHINE

The P-type machine may be regarded as an LCM without a tape, and whose description is largely incomplete. When a configuration is reached, for which the action is undetermined, a random choice for the missing data is made and the appropriate entry is made in the description, tentatively, and is applied. When a pain stimulus occurs all tentative entries are cancelled, and when a pleasure stimulus occurs they are all made permanent.

Specifically. The situation is a number $s = 1, 2, \ldots, N$ and corresponds to the configuration of the incomplete machine. The character is a table of N entries showing the behaviour of the machine in each situation. Each entry has to say something both about the next situation and about what action the machine has to take. The action part may be either

[[9]]   (i) To do some externally visible act A_1 or $A_2 \ldots A_K$

(ii) To set one of the memory units $M_1 \ldots M_R$ either into the '1' condition or into the '0' condition.

The next situation is always the remainder either of $2s$ or of $2s+1$ on division by N. These may be called alternatives 0 and 1. Which alternative applies may be determined by either

(a) one of the memory units

(b) a sense stimulus

(c) the pleasure–pain arrangements.

In each situation it is determined which of these applies when the machine is made, i.e., interference cannot alter which of the three cases applies. Also in cases (a) and (b) interference can have no effect. In case (c) the entry in the character table may be either U ('uncertain'), or T0 (tentative 0), T1, D0 (definite 0) or D1. When the entry in the character for the current situation is U then the alternative is chosen at random, and the entry in the character is changed to T0 or T1 according as 0 or 1 was chosen. If the character entry was T0 or D0 then the alternative is 0 and if it is T1 or D1 then the alternative is 1. The changes in character include the above mentioned change from U to T0 or T1, and a change of every T to D when a pleasure stimulus occurs, changes of T0 and T1 to U when a pain stimulus occurs.

We may imagine the memory units essentially as 'trigger circuits' or switches. The sense stimuli are means by which the teacher communicates 'unemotionally' to the machine, i.e., otherwise than by pleasure and pain stimuli. There are a finite number S of sense stimulus lines, and each always carries either the signal 0 or 1.

A small P-type machine is described in the table below

[[10]]

1	P	A	
2	P	B	M1 = 1
3	P	B	
4	S1	A	M1 = 0
5	M1	C	

18

[[122]]

In this machine there is only one memory unit M1 and one sense line S1. Its behaviour can be described by giving the successive situations together with the actions of the teacher: the latter consist of the values of S1 and the rewards and punishments. At any moment the 'character' consists of the above table with each 'P' replaced by either U, T0, D0 or D1. In working out the behaviour of the machine it is convenient first of all to make up a sequence of random digits for use when the U cases occur. Underneath these we may write the sequence of situations, and have other rows for the corresponding entries from the character, and for the actions of the teacher. The character and the values stored in the memory units may be kept on another sheet. The T entries may be made in pencil and the D entries in ink. A bit of the behaviour of the machine is given below:

Random sequence   0 0 1 1 1 0 0 1 0 0 1 1 0 1 1 0 0 0
Situations   3 1 3 1 3 1 3 1 2 4 4 4 3 2 . .
Alternative given by   U T T T T T U U S S S U T
                       0 0 0 0 0   1 1 1   0
Visible action   B A B A B A B A B A A A B B
Rew. & Pun.   P
Changes in S1   1   0

It will be noticed that the machine very soon got into a repetitive cycle. This became externally visible through the repetitive B A B A B . . . . By means of a pain stimulus this cycle was broken.

It is probably possible to organize these P-type machines into universal machines, but it is not easy because of the form of memory available. It would be necessary to organize the randomly distributed 'memory units' to provide a systematic form of memory, and this would not be easy. If, however, we supply the P-type machine with a systematic external memory this organizing becomes quite feasible. Such a memory could be provided in the form of a tape, and the externally visible operations could include movement to right and left along the tape, and altering the symbol on the tape to 0 or to 1. The sense lines could include one from the symbol on the tape. Alternatively, if the memory were to be finite, e.g., not more than 2^{32} binary digits, we could use a dialling system. (Dialling systems can also be used with an infinite memory, but this is not of much practical interest.) I have succeeded in organizing such a (paper) machine into a universal machine.

The details of the machine involved were as follows. There was a circular memory consisting of 64 squares of which at any moment one was in the machine ('scanned') and motion to right or left were among the 'visible actions'. Changing the symbol on the square was another 'visible action', and the symbol was connected to one of the sense lines S1. The even-numbered squares also had another function, they controlled the dialling of information to or from the main memory. This main memory consisted of 2^{32} binary

19

digits. At any moment one of these digits was connected to the sense line S2. The digit of the main memory concerned was that indicated by the 32 even positioned digits of the circular memory. Another two of the 'visible actions' were printing 0 or 1 in this square of the main memory. There were also three ordinary memory units and three sense units S3, S4, S5. Also six other externally visible actions A,B,C,D,E,F.

This P-type machine with external memory has, it must be admitted, considerably more 'organization' than say the A-type unorganized machine. Nevertheless the fact that it can be organized into a universal machine still remains interesting.

The actual technique by which the 'organizing' of the P-type machine was carried through is perhaps a little disappointing. It is not sufficiently analogous to the kind of process by which a child would really be taught. The process actually adopted was first to let the machine run for a long time with continuous application of pain, and with various changes of the sense data S3, S4, S5. Observation of the sequence of externally visible actions for some thousands of moments made it possible to set up a scheme for identifying the situations, i.e., by which one could at any moment find out what the situation was, except that the situations as a whole had been renamed. A similar investigation, with less use of punishment, enables one to find the situations which are affected by the sense lines; the data about the situations involving the memory units can also be found but with more difficulty. At this stage the character has been reconstructed. There are no occurrences of T0, T1, D0, D1. The next stage is to think up some way of replacing the 0s of the character by D0, D1 in such a way as to give the desired modification. This will normally be possible with the suggested number of situations (1000), memory units, etc. The final stage is the conversion of the character into the chosen one. This may be done simply by allowing the machine to wander at random through a sequence of situations, and applying pain stimuli when the wrong choice is made, pleasure stimuli when the right one is made. It is best also to apply pain stimuli when irrelevant choices are made. This is to prevent getting isolated in a ring of irrelevant situations. The machine is now 'ready for use'.

The form of universal machine actually produced in this process was as follows. Each instruction consisted of 128 digits, which we may regard as forming four sets of 32, each of which describes one place in the main memory. These places may be called P,Q,R,S. The meaning of the instruction is that if p is the digit at P and q that at Q then $1 - pq$ is to be transferred to position R and that the next instruction will be found in the 128 digits beginning at S. This gives a UPCM, though with rather less facilities than are available say on the ACE.

I feel that more should be done on these lines. I would like to investigate other types of unorganized machines, and also to try out organizing methods that would be more nearly analogous to our 'methods of education'. I made

20

a start on the latter but found the work altogether too laborious at present. When some electronic machines are in actual operation I hope that they will make this more feasible. It should be easy to make a model of any particular machine that one wishes to work on within such a UPCM instead of having to work with a paper machine as at present. If also one decided on quite definite 'teaching policies' these could also be programmed into the machine. One would then allow the whole system to run for an appreciable period, and then break in as a kind of 'inspector of schools' and see what progress had been made. One might also be able to make some progress with unorganized machines more like the A- and B-types. The work involved with these is altogether too great for pure paper-machine work.

One particular kind of phenomenon I had been hoping to find in connection with the P-type machines. This was the incorporation of old routines into new. One might have 'taught' (i.e., modified or organized) a machine to add (say). Later one might teach it to multiply by small numbers by repeated addition and so arrange matters that the same set of situations which formed the addition routine, as originally taught, was also used in the additions ⟦11⟧ involved in the multiplication. Although I was able to obtain a fairly detailed picture of how this might happen I was not able to do experiments on a sufficient scale for such phenomena to be seen as part of a large context.

I also hoped to find something rather similar to the 'irregular verbs' which add variety to language. We seem to be quite content that things should not obey too mathematically regular rules. By long experience we can pick up and apply the most complicated rules without being able to enunciate them at all. I rather suspect that a P-type machine without the systematic memory would behave in a rather similar manner because of the randomly distributed memory units. Clearly this could only be verified by very painstaking work; by the very nature of the problem 'mass production' methods like built-in teaching procedures could not help.

DISCIPLINE AND INITIATIVE

If the untrained infant's mind is to become an intelligent one, it must acquire both discipline and initiative. So far we have been considering only discipline. To convert a brain or machine into a universal machine is the extremest form of discipline. Without something of this kind one cannot set up proper communication. But discipline is certainly not enough in itself to produce intelligence. That which is required in addition we call initiative. This statement will have to serve as a definition. Our task is to discover the nature of this residue as it occurs in man, and to try and copy it in machines.

Two possible methods of setting about this present themselves. On the one hand we have fully disciplined machines immediately available, or in a matter of months or years, in the form of various UPCMs. We might try to graft some initiative onto these. This would probably take the form of programming the machine to do every kind of job that could be done, as a

21

matter of principle, whether it were economical to do it by machine or not. Bit by bit one would be able to allow the machine to make more and more 'choices' or 'decisions'. One would eventually find it possible to program it so as to make its behaviour be the logical result of a comparatively small number of general principles. When these became sufficiently general, interference would no longer be necessary, and the machine would have 'grown up'. This may be called the 'direct method'.

The other method is to start with an unorganized machine and to try to bring both discipline and initiative into it at once, i.e., instead of trying to organize the machine to become a universal machine, to organize it for initiative as well. Both methods should, I think, be attempted.

Intellectual, genetical and cultural searches

A very typical sort of problem requiring some sort of initiative consists of those of the form 'Find a number n such that . . .'. This form covers a very great variety of problems. For instance problems of the form 'See if you can find a way of calculating the function which will enable us to obtain the values for arguments . . . to accuracy . . . within a time . . . using the UPCM . . .' are reducible to this form, for the problem is clearly equivalent to that of finding a program to put on the machine in question, and it is easy to put the programs into correspondence with the positive integers in such a way that given either the number or the program the other can easily be found. We should not go far wrong for the time being if we assumed that all problems were reducible to this form. It will be time to think again when something turns up which is obviously not of this form.

The crudest way of dealing with such a problem is to take the integers in order and to test each one to see whether it has the required property, and to go on until one is found which has it. Such a method will only be successful in the simplest cases. For instance in the case of problems of the kind mentioned above, where one is really searching for a program, the number required will normally be somewhere between 2^{1000} and $2^{1,000,000}$. For practical work therefore some more expeditious method is necessary. In a number of cases the following method would be successful. Starting with a UPCM we first put a program into it which corresponds to building in a logical system (like Russell's *Principia Mathematica*). This would not determine the behaviour of the machine completely: at various stages more than one choice as to the next step would be possible. We might arrange, however, to take all possible arrangement of choices in order, and go on until the machine proved a theorem, which, by its form, could be verified to give a solution of the problem. This may be seen to be a conversion of the original problem into another of the same form. Instead of searching through values of the original variable n one searches through values of something else. In practice when solving problems of the above kind one will probably apply some very complex 'transformation' of the original problem, involving searching through

[[12]]

22

various variables, some more analogous to the original one, some more like a 'search through all proofs'. Further research into intelligence of machinery will probably be very greatly concerned with 'searches' of this kind. We may perhaps call such searches 'intellectual searches'. They might very briefly be defined as 'searches carried out by brains for combinations with particular properties'.

It may be of interest to mention two other kinds of search in this connection. There is the genetical or evolutionary search by which a combination of genes is looked for, the criterion being survival value. The remarkable success of this search confirms to some extent the idea that intellectual activity consists mainly of various kinds of search.

The remaining form of search is what I should like to call the 'cultural search'. As I have mentioned, the isolated man does not develop any intellectual power. It is necessary for him to be immersed in an environment of other men, whose techniques he absorbs during the first twenty years of his life. He may then perhaps do a little research of his own and make a very few discoveries which are passed on to other men. From this point of view the search for new techniques must be regarded as carried out by the human community as a whole, rather than by individuals.

INTELLIGENCE AS AN EMOTIONAL CONCEPT

The extent to which we regard something as behaving in an intelligent manner is determined as much by our own state of mind and training as by the properties of the object under consideration. If we are able to explain and predict its behaviour or if there seems to be little underlying plan, we have little temptation to imagine intelligence. With the same object therefore it is possible that one man would consider it as intelligent and another would not; the second man would have found out the rules of its behaviour.

It is possible to do a little experiment on these lines, even at the present stage of knowledge. It is not difficult to devise a paper machine which will play a not very bad game of chess. Now get three men as subjects for the experiment A,B,C. A and C are to be rather poor chess players, B is the operator who works the paper machine. (In order that he should be able to work it fairly fast it is advisable that he be both mathematician and chess player.) Two rooms are used with some arrangement for communicating moves, and a game is played between C and either A or the paper machine. C may find it quite difficult to tell which he is playing. (This is a rather idealized form of an experiment I have actually done.)

[[13]]

REFERENCES

Church, Alonzo (1936) An unsolvable problem of elementary number theory. *Amer. J. of Math.* **58**, 345–63.

Gödel, K. (1931) Über formal unentscheidbare Satze der Principia Mathematica und verwandter Systeme. *Monatshefte fur Math. und Phys.* **38**, 173–89.

Turing, A. M. (1937) On computable numbers with an application to the Entscheidungsproblem. *Proc. London Math. Soc.* **42**, 230–65.

23

Checking a large routine. by Dr. A. Turing.

How can one check a routine in the sense of making sure that it is right? [[1]] [[2]]

In order that the man who checks may not have too difficult a task the programmer should make a number of definite assertions which can be checked individually, and from which the correctness of the whole programme easily follows.

Consider the analogy of checking an addition.   If it is given as:

$$
\begin{array}{r}
1374 \\
5906 \\
6719 \\
4337 \\
7768 \\
\hline
26104
\end{array}
$$

one must check the whole at one sitting, because of the carries.
But if the totals for the various columns are given, as below:

$$
\begin{array}{r}
1374 \\
5906 \\
6719 \\
4337 \\
7768 \\
\hline
3974 \\
2213 \\
\hline
26104
\end{array}
$$

[[3]]

the checker's work is much easier being split up into the checking of the various assertions $3 + 9 + 7 + 3 + 7 = 29$ etc. and the small addition

$$
\begin{array}{r}
3794 \\
2213 \\
\hline
26104
\end{array}
$$

This principle can be applied to the process of checking a large routine but we will illustrate the method by means of a small routine viz. one to obtain n without the use of a multiplier, multiplication being carried out [[4]]
by repeated addition.

At a typical moment of the process we have recorded r and $s\,r$ for some [[5]]
r, s.   We can change $s\,r$ to $(s+1)\,r$ by addition of r.   When $s = r+1$ [[6]]
we can change r to $r+1$ by a transfer. Unfortunately there is no coding
system sufficiently generally known to justify giving the routine for this
process in full, but the flow diagram given in Fig.1 will be sufficient [[7]]
for illustration.

Each 'box of the flow diagram represents a straight sequence of
instructions without changes of control.   The following convention is used:

(i)   a dashed letter indicates the value at the end of the process
       represented by the box:

(ii)   an undashed letter represents the initial value of a quantity.

One cannot equate similar letters appearing in different boxes, but it
is intended that the following identifications be valid throughout

[[129]]

s		content of line 27 of store				
r	"	"	"	28	"	"
n	"	"	"	29	"	"
u	"	"	"	30	"	"
v	"	"	"	31	"	"

[[8]] It is also intended that u be s r or something of the sort e.g. it might be (s+1) r or s r-1 but not e.g. $s^2 + r^2$.

[[9]] In order to assist the checker, the programmer should make assertions about the various states that the machine can reach. These assertions may be tabulated as in fig.2. Assertions are only made for the states when certain particular quantities are in control, corresponding to the ringed letters in the flow diagram. One column of the table is used for each such situation of the control. Other quantities are also needed to specify the condition of the machine completely: in our case it is sufficient to give r and s. The upper part of the table gives the various contents of the store lines in the various conditions of the machine, and restrictions on the

[[10]] quantities s, r (which we may call inductive variables). The lower part tells us which of the conditions will be the next to occur.

  The checker has to verify that the columns corresponding to the initial condition and the stopped condition agree with the claims that are made for the routine as a whole. In this case the claim is that if we start with

[[11]] control in condition D and with n in line 29 we shall find a quantity in

[[12]] [[13]] line 31 when the machine stops which is r (provided this is less than 2^{40}, but this condition has been ignored).

  He has also to verify that each of the assertions in the lower half of the table is correct. In doing this the columns may be taken in any order and quite independently. Thus for column B the checker would argue. "From the flow diagram we see that after B the box $v^1 = u$ applies. From

[[14]] the upper part of the column for B we have u = r . Hence $v^1 = r$ i.e.

[[15]] the entry for v i.e. for line 31 in C should be r . The other entries are the same as in B".

  Finally the checker has to verify that the process comes to an end. Here again he should be assisted by the programmer giving a further definite assertion to be verified. This may take the form of a quantity which is asserted to decrease continually and vanish when the machine stops. To the pure mathematician it is natural to give an ordinal number. In this problem the ordinal might be $(n - r) w^2 + (r - s) w + k$. A less highbrow form of the same thing would be to give the integer $2^{80} (n - r) + 2^{40} (r - s) + k$. Taking the latter case and the step from B to C there would be a decrease

[[16]] from $2^{80} (n - r) + 2^{40} (r - s) + 5$ to $2^{80} (n - v) + 2^{40} (r - s) + 4$. In the step from F to B there is a decrease from $2^{80} (n - r) + 2^{40} (r - s) + 1$

[[17]] to $2^{80} (n - r 1) + 2^{40} (r + 1 - s) + 5$.

  In the course of checking that the process comes to an end the time involved may also be estimated by arranging that the decreasing quantity represents an upper bound to the time till the machine stops.

[[130]]

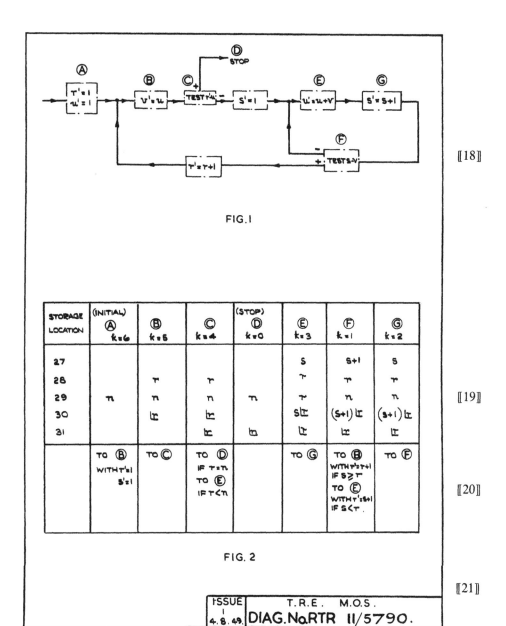

FIG. 1

FIG. 2

ISSUE 1 4.8.49   T.R.E.   M.O.S.   DIAG. No. RTR 11/5790.

[[18]]

[[19]]

[[20]]

[[21]]

[[131]]

VOL. LIX.   No. 236.]   [October, 1950

MIND

A QUARTERLY REVIEW

OF

PSYCHOLOGY AND PHILOSOPHY

I.—COMPUTING MACHINERY AND INTELLIGENCE

By A. M. TURING

1. *The Imitation Game.*

I PROPOSE to consider the question, 'Can machines think?' This should begin with definitions of the meaning of the terms 'machine' and 'think'. The definitions might be framed so as to reflect so far as possible the normal use of the words, but this attitude is dangerous. If the meaning of the words 'machine' and 'think' are to be found by examining how they are commonly used it is difficult to escape the conclusion that the meaning and the answer to the question, 'Can machines think?' is to be sought in a statistical survey such as a Gallup poll. But this is absurd. Instead of attempting such a definition I shall replace the question by another, which is closely related to it and is expressed in relatively unambiguous words.

The new form of the problem can be described in terms of a game which we call the 'imitation game'. It is played with three people, a man (A), a woman (B), and an interrogator (C) who may be of either sex. The interrogator stays in a room apart from the other two. The object of the game for the interrogator is to determine which of the other two is the man and which is the woman. He knows them by labels X and Y, and at the end of the game he says either 'X is A and Y is B' or 'X is B and Y is A'. The interrogator is allowed to put questions to A and B thus:

C: Will X please tell me the length of his or her hair?

Now suppose X is actually A, then A must answer. It is A's

433

object in the game to try and cause C to make the wrong identification.　His answer might therefore be

'My hair is shingled, and the longest strands are about nine inches long.'

In order that tones of voice may not help the interrogator the answers should be written, or better still, typewritten.　The ideal arrangement is to have a teleprinter communicating between the two rooms.　Alternatively the question and answers can be repeated by an intermediary.　The object of the game for the third player (B) is to help the interrogator.　The best strategy for her is probably to give truthful answers.　She can add such things as ' I am the woman, don't listen to him ! ' to her answers, but it will avail nothing as the man can make similar remarks.

[[1]]

We now ask the question, ' What will happen when a machine takes the part of A in this game ? '　Will the interrogator decide wrongly as often when the game is played like this as he does when the game is played between a man and a woman ?　These questions replace our original, ' Can machines think ? '

2. *Critique of the New Problem.*

As well as asking, ' What is the answer to this new form of the question ', one may ask, ' Is this new question a worthy one to investigate ? '　This latter question we investigate without further ado, thereby cutting short an infinite regress.

The new problem has the advantage of drawing a fairly sharp line between the physical and the intellectual capacities of a man. No engineer or chemist claims to be able to produce a material which is indistinguishable from the human skin.　It is possible that at some time this might be done, but even supposing this invention available we should feel there was little point in trying to make a ' thinking machine ' more human by dressing it up in such artificial flesh.　The form in which we have set the problem reflects this fact in the condition which prevents the interrogator from seeing or touching the other competitors, or hearing their voices.　Some other advantages of the proposed criterion may be shown up by specimen questions and answers.　Thus :

Q : Please write me a sonnet on the subject of the Forth Bridge.

A : Count me out on this one.　I never could write poetry.

Q : Add 34957 to 70764

A : (Pause about 30 seconds and then give as answer) 105621.

Q : Do you play chess ?

A : Yes.

Q : I have K at my K1, and no other pieces. You have only
     K at K6 and R at R1. It is your move. What do you
     play ?
A : (After a pause of 15 seconds) R-R8 mate.

The question and answer method seems to be suitable for
introducing almost any one of the fields of human endeavour that
we wish to include. We do not wish to penalise the machine
for its inability to shine in beauty competitions, nor to penalise
a man for losing in a race against an aeroplane. The conditions
of our game make these disabilities irrelevant. The ' witnesses '
can brag, if they consider it advisable, as much as they please
about their charms, strength or heroism, but the interrogator
cannot demand practical demonstrations.

The game may perhaps be criticised on the ground that the
odds are weighted too heavily against the machine. If the man
were to try and pretend to be the machine he would clearly make
a very poor showing. He would be given away at once by slowness
and inaccuracy in arithmetic. May not machines carry out something which ought to be described as thinking but which is very
different from what a man does ? This objection is a very strong
one, but at least we can say that if, nevertheless, a machine can
be constructed to play the imitation game satisfactorily, we need
not be troubled by this objection.

It might be urged that when playing the ' imitation game '
the best strategy for the machine may possibly be something
other than imitation of the behaviour of a man. This may be, but
I think it is unlikely that there is any great effect of this kind.
In any case there is no intention to investigate here the theory
of the game, and it will be assumed that the best strategy is
to try to provide answers that would naturally be given by a man.

3. *The Machines concerned in the Game.*

The question which we put in § 1 will not be quite definite
until we have specified what we mean by the word ' machine '.
It is natural that we should wish to permit every kind of engineering
technique to be used in our machines. We also wish to allow the
possibility than an engineer or team of engineers may construct
a machine which works, but whose manner of operation cannot
be satisfactorily described by its constructors because they have
applied a method which is largely experimental. Finally, we
wish to exclude from the machines men born in the usual manner.
It is difficult to frame the definitions so as to satisfy these three
conditions. One might for instance insist that the team of

[[2]]

engineers should be all of one sex, but this would not really be satisfactory, for it is probably possible to rear a complete individual from a single cell of the skin (say) of a man. To do so would be a feat of biological technique deserving of the very highest praise, but we would not be inclined to regard it as a case of 'constructing a thinking machine'. This prompts us to abandon the requirement that every kind of technique should be permitted. We are the more ready to do so in view of the fact that the present interest in 'thinking machines' has been aroused by a particular kind of machine, usually called an 'electronic computer' or 'digital computer'. Following this suggestion we only permit digital computers to take part in our game.

This restriction appears at first sight to be a very drastic one. I shall attempt to show that it is not so in reality. To do this necessitates a short account of the nature and properties of these computers.

It may also be said that this identification of machines with digital computers, like our criterion for 'thinking', will only be unsatisfactory if (contrary to my belief), it turns out that digital computers are unable to give a good showing in the game.

There are already a number of digital computers in working order, and it may be asked, 'Why not try the experiment straight away? It would be easy to satisfy the conditions of the game. A number of interrogators could be used, and statistics compiled to show how often the right identification was given.' The short answer is that we are not asking whether all digital computers would do well in the game nor whether the computers at present available would do well, but whether there are imaginable computers which would do well. But this is only the short answer. We shall see this question in a different light later.

4. *Digital Computers.*

The idea behind digital computers may be explained by saying that these machines are intended to carry out any operations which could be done by a human computer. The human computer is supposed to be following fixed rules; he has no authority to deviate from them in any detail. We may suppose that these rules are supplied in a book, which is altered whenever he is put on to a new job. He has also an unlimited supply of paper on which he does his calculations. He may also do his multiplications and additions on a 'desk machine', but this is not important.

If we use the above explanation as a definition we shall be in

[[3]]

danger of circularity of argument. We avoid this by giving an outline of the means by which the desired effect is achieved. A digital computer can usually be regarded as consisting of three parts :

> (i) Store.
> (ii) Executive unit.
> (iii) Control.

The store is a store of information, and corresponds to the human computer's paper, whether this is the paper on which he does his calculations or that on which his book of rules is printed. In so far as the human computer does calculations in his head a part of the store will correspond to his memory.

The executive unit is the part which carries out the various individual operations involved in a calculation. What these individual operations are will vary from machine to machine. Usually fairly lengthy operations can be done such as 'Multiply 3540675445 by 7076345687' but in some machines only very simple ones such as 'Write down 0' are possible.

We have mentioned that the 'book of rules' supplied to the computer is replaced in the machine by a part of the store. It is then called the 'table of instructions'. It is the duty of the control to see that these instructions are obeyed correctly and in the right order. The control is so constructed that this necessarily happens.

The information in the store is usually broken up into packets of moderately small size. In one machine, for instance, a packet might consist of ten decimal digits. Numbers are assigned to the parts of the store in which the various packets of information are stored, in some systematic manner. A typical instruction might say—

'Add the number stored in position 6809 to that in 4302 and put the result back into the latter storage position'.

Needless to say it would not occur in the machine expressed in English. It would more likely be coded in a form such as 6809430217. Here 17 says which of various possible operations is to be performed on the two numbers. In this case the operation is that described above, viz. 'Add the number. . . .' It will be noticed that the instruction takes up 10 digits and so forms one packet of information, very conveniently. The control will normally take the instructions to be obeyed in the order of the positions in which they are stored, but occasionally an instruction such as

⟦4⟧

' Now obey the instruction stored in position 5606, and continue from there '

may be encountered, or again

' If position 4505 contains 0 obey next the instruction stored in 6707, otherwise continue straight on.'

Instructions of these latter types are very important because they make it possible for a sequence of operations to be repeated over and over again until some condition is fulfilled, but in doing so to obey, not fresh instructions on each repetition, but the same ones over and over again. To take a domestic analogy. Suppose Mother wants Tommy to call at the cobbler's every morning on his way to school to see if her shoes are done, she can ask him afresh every morning. Alternatively she can stick up a notice once and for all in the hall which he will see when he leaves for school and which tells him to call for the shoes, and also to destroy the notice when he comes back if he has the shoes with him.

The reader must accept it as a fact that digital computers can be constructed, and indeed have been constructed, according to the principles we have described, and that they can in fact mimic the actions of a human computer very closely.

The book of rules which we have described our human computer as using is of course a convenient fiction. Actual human computers really remember what they have got to do. If one wants to make a machine mimic the behaviour of the human computer in some complex operation one has to ask him how it is done, and then translate the answer into the form of an instruction table. Constructing instruction tables is usually described as ' programming '. To ' programme a machine to carry out the operation A ' means to put the appropriate instruction table into the machine so that it will do A.

[[5]]   An interesting variant on the idea of a digital computer is a 'digital computer with a random element'. These have instructions involving the throwing of a die or some equivalent electronic process ; one such instruction might for instance be,' Throw the die and put the resulting number into store 1000 '. Sometimes such a machine is described as having free will (though I would not use this phrase myself). It is not normally possible to determine from observing a machine whether it has a random element, for a similar effect can be produced by such devices as making the choices depend on the digits of the decimal for π.

Most actual digital computers have only a finite store. There is no theoretical difficulty in the idea of a computer with an unlimited store. Of course only a finite part can have been used at any one time. Likewise only a finite amount can have been constructed, but we can imagine more and more being added as required. Such computers have special theoretical interest and will be called infinitive capacity computers.

The idea of a digital computer is an old one. Charles Babbage, Lucasian Professor of Mathematics at Cambridge from 1828 to 1839, planned such a machine, called the Analytical Engine, but it was never completed. Although Babbage had all the essential ideas, his machine was not at that time such a very attractive prospect. The speed which would have been available would be definitely faster than a human computer but something like 100 times slower than the Manchester machine, itself one of the slower of the modern machines. The storage was to be purely mechanical, using wheels and cards.

The fact that Babbage's Analytical Engine was to be entirely mechanical will help us to rid ourselves of a superstition. Importance is often attached to the fact that modern digital computers are electrical, and that the nervous system also is electrical. Since Babbage's machine was not electrical, and since all digital computers are in a sense equivalent, we see that this use of electricity cannot be of theoretical importance. Of course electricity usually comes in where fast signalling is concerned, so that it is not surprising that we find it in both these connections. In the nervous system chemical phenomena are at least as important as electrical. In certain computers the storage system is mainly acoustic. The feature of using electricity is thus seen to be only a very superficial similarity. If we wish to find such similarities we should look rather for mathematical analogies of function.

⟦6⟧

5. *Universality of Digital Computers.*

The digital computers considered in the last section may be classified amongst the 'discrete state machines'. These are the machines which move by sudden jumps or clicks from one quite definite state to another. These states are sufficiently different for the possibility of confusion between them to be ignored. Strictly speaking there are no such machines. Everything really moves continuously. But there are many kinds of machine which can profitably be *thought of* as being discrete state machines. For instance in considering the switches for a lighting system it is a convenient fiction that each switch must be definitely on or definitely off. There must be intermediate positions, but for most purposes we can forget about them. As an example of a discrete state machine we might consider a wheel which clicks

round through 120° once a second, but may be stopped by a
lever which can be operated from outside ; in addition a lamp is
to light in one of the positions of the wheel. This machine could
be described abstractly as follows. The internal state of the
machine (which is described by the position of the wheel) may be
q_1, q_2 or q_3. There is an input signal i_0 or i_1 (position of lever).
The internal state at any moment is determined by the last state
and input signal according to the table

		Last State		
		q_1	q_2	q_3
Input	i_0	q_2	q_3	q_1
	i_1	q_1	q_2	q_3

The output signals, the only externally visible indication of
the internal state (the light) are described by the table

State	q_1	q_2	q_3
Output	o_0	o_0	o_1

This example is typical of discrete state machines. They can be
described by such tables provided they have only a finite number
of possible states.

It will seem that given the initial state of the machine and
the input signals it is always possible to predict all future states.
This is reminiscent of Laplace's view that from the complete
state of the universe at one moment of time, as described by the
positions and velocities of all particles, it should be possible to
predict all future states. The prediction which we are considering
is, however, rather nearer to practicability than that considered
by Laplace. The system of the ' universe as a whole ' is such
that quite small errors in the initial conditions can have an
overwhelming effect at a later time. The displacement of a
single electron by a billionth of a centimetre at one moment
might make the difference between a man being killed by an
avalanche a year later, or escaping. It is an essential property
of the mechanical systems which we have called ' discrete state
machines ' that this phenomenon does not occur. Even when we
consider the actual physical machines instead of the idealised
machines, reasonably accurate knowledge of the state at one
moment yields reasonably accurate knowledge any number of
steps later.

As we have mentioned, digital computers fall within the class of discrete state machines. But the number of states of which such a machine is capable is usually enormously large. For instance, the number for the machine now working at Manchester it about $2^{165,000}$, *i.e.* about $10^{50,000}$. Compare this with our example of the clicking wheel described above, which had three states. It is not difficult to see why the number of states should be so immense. The computer includes a store corresponding to the paper used by a human computer. It must be possible to write into the store any one of the combinations of symbols which might have been written on the paper. For simplicity suppose that only digits from 0 to 9 are used as symbols. Variations in handwriting are ignored. Suppose the computer is allowed 100 sheets of paper each containing 50 lines each with room for 30 digits. Then the number of states is $10^{100 \times 50 \times 30}$, *i.e.* $10^{150,000}$. This is about the number of states of three Manchester machines put together. The logarithm to the base two of the number of states is usually called the 'storage capacity' of the machine. Thus the Manchester machine has a storage capacity of about 165,000 and the wheel machine of our example about 1·6. If two machines are put together their capacities must be added to obtain the capacity of the resultant machine. This leads to the possibility of statements such as 'The Manchester machine contains 64 magnetic tracks each with a capacity of 2560, eight electronic tubes with a capacity of 1280. Miscellaneous storage amounts to about 300 making a total of 174,380.'

Given the table corresponding to a discrete state machine it is possible to predict what it will do. There is no reason why this calculation should not be carried out by means of a digital computer. Provided it could be carried out sufficiently quickly the digital computer could mimic the behaviour of any discrete state machine. The imitation game could then be played with the machine in question (as B) and the mimicking digital computer (as A) and the interrogator would be unable to distinguish them. Of course the digital computer must have an adequate storage capacity as well as working sufficiently fast. Moreover, it must be programmed afresh for each new machine which it is desired to mimic.

This special property of digital computers, that they can mimic any discrete state machine, is described by saying that they are *universal* machines. The existence of machines with this property has the important consequence that, considerations of speed apart, it is unnecessary to design various new machines to do various computing processes. They can all be

[[7]]

done with one digital computer, suitably programmed for each case. It will be seen that as a consequence of this all digital computers are in a sense equivalent.

We may now consider again the point raised at the end of §3. It was suggested tentatively that the question, 'Can machines think?' should be replaced by 'Are there imaginable digital computers which would do well in the imitation game?' If we wish we can make this superficially more general and ask 'Are there discrete state machines which would do well?' But in view of the universality property we see that either of these questions is equivalent to this, 'Let us fix our attention on one particular digital computer C. Is it true that by modifying this computer to have an adequate storage, suitably increasing its speed of action, and providing it with an appropriate programme, C can be made to play satisfactorily the part of A in the imitation game, the part of B being taken by a man?'

6. *Contrary Views on the Main Question.*

We may now consider the ground to have been cleared and we are ready to proceed to the debate on our question, 'Can machines think?' and the variant of it quoted at the end of the last section. We cannot altogether abandon the original form of the problem, for opinions will differ as to the appropriateness of the substitution and we must at least listen to what has to be said in this connexion.

It will simplify matters for the reader if I explain first my own beliefs in the matter. Consider first the more accurate form of the question. I believe that in about fifty years' time it will be possible to programme computers, with a storage capacity of about 10^9, to make them play the imitation game so well that an average interrogator will not have more than 70 per cent. chance of making the right identification after five minutes of questioning. The original question, 'Can machines think?' I believe to be too meaningless to deserve discussion. Nevertheless I believe that at the end of the century the use of words and general educated opinion will have altered so much that one will be able to speak of machines thinking without expecting to be contradicted. I believe further that no useful purpose is served by concealing these beliefs. The popular view that scientists proceed inexorably from well-established fact to well-established fact, never being influenced by any unproved conjecture, is quite mistaken. Provided it is made clear which are proved facts and which are conjectures, no harm can result. Conjectures are of great importance since they suggest useful lines of research.

[[8]]

I now proceed to consider opinions opposed to my own.

(1) *The Theological Objection.* Thinking is a function of man's immortal soul. God has given an immortal soul to every man and woman, but not to any other animal or to machines. Hence no animal or machine can think.

I am unable to accept any part of this, but will attempt to reply in theological terms. I should find the argument more convincing if animals were classed with men, for there is a greater difference, to my mind, between the typical animate and the inanimate than there is between man and the other animals. The arbitrary character of the orthodox view becomes clearer if we consider how it might appear to a member of some other religious community. How do Christians regard the Moslem view that women have no souls ? But let us leave this point aside and return to the main argument. It appears to me that the argument quoted above implies a serious restriction of the omnipotence of the Almighty. It is admitted that there are certain things that He cannot do such as making one equal to two, but should we not believe that He has freedom to confer a soul on an elephant if He sees fit ? We might expect that He would only exercise this power in conjunction with a mutation which provided the elephant with an appropriately improved brain to minister to the needs of this soul. An argument of exactly similar form may be made for the case of machines. It may seem different because it is more difficult to " swallow ". But this really only means that we think it would be less likely that He would consider the circumstances suitable for conferring a soul. The circumstances in question are discussed in the rest of this paper. In attempting to construct such machines we should not be irreverently usurping His power of creating souls, any more than we are in the procreation of children : rather we are, in either case, instruments of His will providing mansions for the souls that He creates.

However, this is mere speculation. I am not very impressed with theological arguments whatever they may be used to support. Such arguments have often been found unsatisfactory in the past. In the time of Galileo it was argued that the texts, " And the sun stood still . . . and hasted not to go down about a whole day " (Joshua x. 13) and " He laid the foundations of the earth,

---

[1] Possibly this view is heretical. St. Thomas Aquinas (*Summa Theologica*. quoted by Bertrand Russell, p. 480) states that God cannot make a man to have no soul. But this may not be a real restriction on His powers, but only a result of the fact that men's souls are immortal, and therefore indestructible.

that it should not move at any time " (Psalm cv. 5) were an
adequate refutation of the Copernican theory. With our present
knowledge such an argument appears futile. When that knowledge was not available it made a quite different impression.

(2) *The 'Heads in the Sand' Objection.* "The consequences
of machines thinking would be too dreadful. Let us hope and
believe that they cannot do so."

This argument is seldom expressed quite so openly as in the
form above. But it affects most of us who think about it at all.
We like to believe that Man is in some subtle way superior to the
rest of creation. It is best if he can be shown to be *necessarily*
superior, for then there is no danger of him losing his commanding
position. The popularity of the theological argument is clearly
connected with this feeling. It is likely to be quite strong in intellectual people, since they value the power of thinking more
highly than others, and are more inclined to base their belief
in the superiority of Man on this power.

I do not think that this argument is sufficiently substantial
to require refutation. Consolation would be more appropriate :
perhaps this should be sought in the transmigration of souls.

(3) *The Mathematical Objection.* There are a number of results
of mathematical logic which can be used to show that there
are limitations to the powers of discrete-state machines. The
best known of these results is known as Gödel's theorem,[1] and
shows that in any sufficiently powerful logical system statements
can be formulated which can neither be proved nor disproved
within the system, unless possibly the system itself is inconsistent.
There are other, in some respects similar, results due to *Church*,
Kleene, *Rosser*, and *Turing*. The latter result is the most convenient to consider, since it refers directly to machines, whereas
the others can only be used in a comparatively indirect argument :
for instance if Gödel's theorem is to be used we need in addition
to have some means of describing logical systems in terms of
machines, and machines in terms of logical systems. The result in
question refers to a type of machine which is essentially a digital
computer with an infinite capacity. It states that there are
certain things that such a machine cannot do. If it is rigged up to
give answers to questions as in the imitation game, there will be
some questions to which it will either give a wrong answer, or fail
to give an answer at all however much time is allowed for a reply.
There may, of course, be many such questions, and questions
which cannot be answered by one machine may be satisfactorily

---

[1] Author's names in italics refer to the Bibliography.

answered by another.   We are of course supposing for the present
that the questions are of the kind to which an answer ' Yes '
or ' No ' is appropriate, rather than questions such as ' What do
you think of Picasso ? '   The questions that we know the
machines must fail on are of this type, " Consider the machine
specified as follows. . . . Will this machine ever answer ' Yes '
to any question ? "   The dots are to be replaced by a description of some machine in a standard form, which could be
something like that used in § 5.   When the machine described
bears a certain comparatively simple relation to the machine
which is under interrogation, it can be shown that the answer
is either wrong or not forthcoming.   This is the mathematical
result : it is argued that it proves a disability of machines to
which the human intellect is not subject.

The short answer to this argument is that although it is
established that there are limitations to the powers of any
particular machine, it has only been stated, without any sort
of proof, that no such limitations apply to the human intellect.
But I do not think this view can be dismissed quite so lightly.
Whenever one of these machines is asked the appropriate
critical question, and gives a definite answer, we know that this
answer must be wrong, and this gives us a certain feeling of
superiority.   Is this feeling illusory ?   It is no doubt quite
genuine, but I do not think too much importance should be
attached to it.   We too often give wrong answers to questions
ourselves to be justified in being very pleased at such evidence of
fallibility on the part of the machines.   Further, our superiority
can only be felt on such an occasion in relation to the one machine
over which we have scored our petty triumph.   There would
be no question of triumphing simultaneously over *all* machines.
In short, then, there might be men cleverer than any given
machine, but then again there might be other machines cleverer
again, and so on.

Those who hold to the mathematical argument would, I think,
mostly be willing to accept the imitation game as a basis for
discussion.   Those who believe in the two previous objections
would probably not be interested in any criteria.

(4) *The Argument from Consciousness.*   This argument is very
well expressed in *Professor Jefferson's* Lister Oration for 1949,
from which I quote.   " Not until a machine can write a sonnet
or compose a concerto because of thoughts and emotions felt,
and not by the chance fall of symbols, could we agree that
machine equals brain—that is, not only write it but know that
it had written it.   No mechanism could feel (and not merely

artificially signal, an easy contrivance) pleasure at its successes, grief when its valves fuse, be warmed by flattery, be made miserable by its mistakes, be charmed by sex, be angry or depressed when it cannot get what it wants.''

This argument appears to be a denial of the validity of our test. According to the most extreme form of this view the only way by which one could be sure that a machine thinks is to *be* the machine and to feel oneself thinking. One could then describe these feelings to the world, but of course no one would be justified in taking any notice. Likewise according to this view the only way to know that a *man* thinks is to be that particular man. It is in fact the solipsist point of view. It may be the most logical view to hold but it makes communication of ideas difficult. A is liable to believe ' A thinks but B does not ' whilst B believes ' B thinks but A does not '. Instead of arguing continually over this point it is usual to have the polite convention that everyone thinks.

I am sure that Professor Jefferson does not wish to adopt the extreme and solipsist point of view. Probably he would be quite willing to accept the imitation game as a test. The game (with the player B omitted) is frequently used in practice under the name of *viva voce* to discover whether some one really understands something or has ' learnt it parrot fashion '. Let us listen in to a part of such a *viva voce* :

Interrogator : In the first line of your sonnet which reads ' Shall I compare thee to a summer's day ', would not ' a spring day ' do as well or better ?

Witness : It wouldn't scan.

Interrogator : How about ' a winter's day ' That would scan all right.

Witness : Yes, but nobody wants to be compared to a winter's day.

Interrogator : Would you say Mr. Pickwick reminded you of Christmas ?

Witness : In a way.

Interrogator : Yet Christmas is a winter's day, and I do not think Mr. Pickwick would mind the comparison.

Witness : I don't think you're serious. By a winter's day one means a typical winter's day, rather than a special one like Christmas.

And so on. What would Professor Jefferson say if the sonnet-writing machine was able to answer like this in the *viva voce*? I do not know whether he would regard the machine as ' merely artificially signalling ' these answers, but if the answers were as satisfactory and sustained as in the above passage I do not think he would describe it as 'an easy contrivance '. This phrase is, I think, intended to cover such devices as the inclusion in the machine of a record of someone reading a sonnet, with appropriate switching to turn it on from time to time.

In short then, I think that most of those who support the argument from consciousness could be persuaded to abandon it rather than be forced into the solipsist position. They will then probably be willing to accept our test.

I do not wish to give the impression that I think there is no mystery about consciousness. There is, for instance, something of a paradox connected with any attempt to localise it. But I do not think these mysteries necessarily need to be solved before we can answer the question with which we are concerned in this paper.

(5) *Arguments from Various Disabilities.* These arguments take the form, " I grant you that you can make machines do all the things you have mentioned but you will never be able to make one to do X ". Numerous features X are suggested in this connexion. I offer a selection :

Be kind, resourceful, beautiful, friendly (p. 448), have initiative, have a sense of humour, tell right from wrong, make mistakes (p. 448), fall in love, enjoy strawberries and cream (p. 448), make some one fall in love with it, learn from experience (pp. 456 f.), use words properly, be the subject of its own thought (p. 449), have as much diversity of behaviour as a man, do something really new (p. 450). (Some of these disabilities are given special consideration as indicated by the page numbers.)

No support is usually offered for these statements. I believe they are mostly founded on the principle of scientific induction. A man has seen thousands of machines in his lifetime. From what he sees of them he draws a number of general conclusions. They are ugly, each is designed for a very limited purpose, when required for a minutely different purpose they are useless, the variety of behaviour of any one of them is very small, etc., etc. Naturally he concludes that these are necessary properties of machines in general. Many of these limitations are associated with the very small storage capacity of most machines. (I am assuming that the idea of storage capacity is extended in some way to cover machines other than discrete-state machines.

The exact definition does not matter as no mathematical accuracy is claimed in the present discussion.) A few years ago, when very little had been heard of digital computers, it was possible to elicit much incredulity concerning them, if one mentioned their properties without describing their construction. That was presumably due to a similar application of the principle of scientific induction. These applications of the principle are of course largely unconscious. When a burnt child fears the fire and shows that he fears it by avoiding it, I should say that he was applying scientific induction. (I could of course also describe his behaviour in many other ways.) The works and customs of mankind do not seem to be very suitable material to which to apply scientific induction. A very large part of space-time must be investigated, if reliable results are to be obtained. Otherwise we may (as most English children do) decide that everybody speaks English, and that it is silly to learn French.

There are, however, special remarks to be made about many of the disabilities that have been mentioned. The inability to enjoy strawberries and cream may have struck the reader as frivolous. Possibly a machine might be made to enjoy this delicious dish, but any attempt to make one do so would be idiotic. What is important about this disability is that it contributes to some of the other disabilities, *e.g.* to the difficulty of the same kind of friendliness occurring between man and machine as between white man and white man, or between black man and black man.

The claim that " machines cannot make mistakes " seems a curious one. One is tempted to retort, " Are they any the worse for that ? " But let us adopt a more sympathetic attitude, and try to see what is really meant. I think this criticism can be explained in terms of the imitation game. It is claimed that the interrogator could distinguish the machine from the man simply by setting them a number of problems in arithmetic. The machine would be unmasked because of its deadly accuracy. The reply to this is simple. The machine (programmed for playing the game) would not attempt to give the *right* answers to the arithmetic problems. It would deliberately introduce [[9]] mistakes in a manner calculated to confuse the interrogator. A mechanical fault would probably show itself through an unsuitable decision as to what sort of a mistake to make in the arithmetic. Even this interpretation of the criticism is not sufficiently sympathetic. But we cannot afford the space to go into it much further. It seems to me that this criticism depends on a confusion between two kinds of mistake. We may call them ' errors of functioning ' and ' errors of conclusion '. Errors of functioning are due to some mechanical or electrical fault which causes the machine to behave otherwise than it was designed to do. In philosophical discussions one likes to ignore the possibility of such errors ; one is therefore discussing ' abstract machines '. These abstract machines are mathematical fictions rather than physical objects. By definition they are incapable of errors of functioning. In this sense we can truly say that ' machines can never make mistakes '. Errors of conclusion can only arise when some meaning is attached to the output signals from the machine. The machine might, for instance, type out mathematical equations, or sentences in English. When a false proposition is typed we say that the machine has committed an error of conclusion. There is clearly no reason at all for saying that a machine cannot make this kind of mistake. It might do nothing but type out repeatedly ' $0 = 1$ '. To take a less perverse example, it might have some method for drawing conclusions by scientific induction. We must expect such a method to lead occasionally to erroneous results.

The claim that a machine cannot be the subject of its own thought can of course only be answered if it can be shown that the machine has *some* thought with *some* subject matter. Nevertheless, ' the subject matter of a machine's operations ' does seem to mean something, at least to the people who deal with it. If, for instance, the machine was trying to find a solution of the equation $x^2 - 40x - 11 = 0$ one would be tempted to describe this equation as part of the machine's subject matter at that moment. In this sort of sense a machine undoubtedly can be its own subject matter. It may be used to help in making up its own programmes, or to predict the effect of alterations in its own structure. By observing the results of its own behaviour it can modify its own programmes so as to achieve some purpose more effectively. These are possibilities of the near future, rather than Utopian dreams.

The criticism that a machine cannot have much diversity of behaviour is just a way of saying that it cannot have much storage capacity. Until fairly recently a storage capacity of even a thousand digits was very rare.

The criticisms that we are considering here are often disguised forms of the argument from consciousness. Usually if one maintains that a machine *can* do one of these things, and describes the kind of method that the machine could use, one will not make

[[10]]

much of an impression.   It is thought that the method (whatever
it may be, for it must be mechanical) is really rather base.
Compare the parenthesis in Jefferson's statement quoted on p. 21.

(6) *Lady Lovelace's Objection.*   Our most detailed information
of Babbage's Analytical Engine comes from a memoir by *Lady
Lovelace.*   In it she states, "The Analytical Engine has no pre-
tensions to *originate* anything.   It can do *whatever we know how
to order it* to perform " (her italics).   This statement is quoted
by *Hartree* (p. 70) who adds : "This does not imply that it
may not be possible to construct electronic equipment which
will 'think for itself ', or in which, in biological terms, one could
set up a conditioned reflex, which would serve as a basis for
'learning '.   Whether this is possible in principle or not is a
stimulating and exciting question, suggested by some of these
recent developments.   But it did not seem that the machines
constructed or projected at the time had this property ".

I am in thorough agreement with Hartree over this.   It will
be noticed that he does not assert that the machines in question
had not got the property, but rather that the evidence available
to Lady Lovelace did not encourage her to believe that they had it.
It is quite possible that the machines in question had in a sense
got this property.   For suppose that some discrete-state machine
has the property.   The Analytical Engine was a universal
digital computer, so that, if its storage capacity and speed were
adequate, it could by suitable programming be made to mimic
the machine in question.   Probably this argument did not
occur to the Countess or to Babbage.   In any case there was no
obligation on them to claim all that could be claimed.

This whole question will be considered again under the heading
of learning machines.

A variant of Lady Lovelace's objection states that a machine
can 'never do anything really new '.   This may be parried for a
moment with the saw, 'There is nothing new under the sun '.
Who can be certain that 'original work ' that he has done was
not simply the growth of the seed planted in him by teaching,
or the effect of following well-known general principles.   A
better variant of the objection says that a machine can never
'take us by surprise '.   This statement is a more direct challenge
and can be met directly.   Machines take me by surprise with
great frequency.   This is largely because I do not do sufficient
calculation to decide what to expect them to do, or rather because,
although I do a calculation, I do it in a hurried, slipshod fashion,
taking risks.   Perhaps I say to myself, 'I suppose the voltage
here ought to be the same as there : anyway let's assume it is '

Naturally I am often wrong, and the result is a surprise for me for by the time the experiment is done these assumptions have been forgotten. These admissions lay me open to lectures on the subject of my vicious ways, but do not throw any doubt on my credibility when I testify to the surprises I experience.

I do not expect this reply to silence my critic. He will probably say that such surprises are due to some creative mental act on my part, and reflect no credit on the machine. This leads us back to the argument from consciousness, and far from the idea of surprise. It is a line of argument we must consider closed, but it is perhaps worth remarking that the appreciation of something as surprising requires as much of a ' creative mental act ' whether the surprising event originates from a man, a book, a machine or anything else.

The view that machines cannot give rise to surprises is due, I believe, to a fallacy to which philosophers and mathematicians are particularly subject. This is the assumption that as soon as a fact is presented to a mind all consequences of that fact spring into the mind simultaneously with it. It is a very useful assumption under many circumstances, but one too easily forgets that it is false. A natural consequence of doing so is that one then assumes that there is no virtue in the mere working out of consequences from data and general principles.

(7) *Argument from Continuity in the Nervous System.* The nervous system is certainly not a discrete-state machine. A small error in the information about the size of a nervous impulse impinging on a neuron, may make a large difference to the size of the outgoing impulse. It may be argued that, this being so, one cannot expect to be able to mimic the behaviour of the nervous system with a discrete-state system.

It is true that a discrete-state machine must be different from a continuous machine. But if we adhere to the conditions of the imitation game, the interrogator will not be able to take any advantage of this difference. The situation can be made clearer if we consider some other simpler continuous machine. A differential analyser will do very well. (A differential analyser is a certain kind of machine not of the discrete-state type used for some kinds of calculation.) Some of these provide their answers in a typed form, and so are suitable for taking part in the game. It would not be possible for a digital computer to predict exactly what answers the differential analyser would give to a problem, but it would be quite capable of giving the right sort of answer. For instance, if asked to give the value of π (actually about 3·1416) it would be reasonable

[[11]]

to choose at random between the values 3·12, 3·13, 3·14, 3·15, 3·16 with the probabilities of 0·05, 0·15, 0·55, 0·19, 0·06 (say). Under these circumstances it would be very difficult for the interrogator to distinguish the differential analyser from the digital computer.

(8) *The Argument from Informality of Behaviour.* It is not possible to produce a set of rules purporting to describe what a man should do in every conceivable set of circumstances. One might for instance have a rule that one is to stop when one sees a red traffic light, and to go if one sees a green one, but what if by some fault both appear together? One may perhaps decide that it is safest to stop. But some further difficulty may well arise from this decision later. To attempt to provide rules of conduct to cover every eventuality, even those arising from traffic lights, appears to be impossible. With all this I agree.

[[12]]

From this it is argued that we cannot be machines. I shall try to reproduce the argument, but I fear I shall hardly do it justice. It seems to run something like this. ' If each man had a definite set of rules of conduct by which he regulated his life he would be no better than a machine. But there are no such rules, so men cannot be machines.' The undistributed middle is glaring. I do not think the argument is ever put quite like this, but I believe this is the argument used nevertheless. There may however be a certain confusion between ' rules of conduct ' and ' laws of behaviour ' to cloud the issue. By ' rules of conduct ' I mean precepts such as ' Stop if you see red lights ', on which one can act, and of which one can be conscious. By ' laws of behaviour ' I mean laws of nature as applied to a man's body such as ' if you pinch him he will squeak '. If we substitute ' laws of behaviour which regulate his life ' for ' laws of conduct by which he regulates his life ' in the argument quoted the undistributed middle is no longer insuperable. For we believe that it is not only true that being regulated by laws of behaviour implies being some sort of machine (though not necessarily a discrete-state machine), but that conversely being such a machine implies being regulated by such laws. However, we cannot so easily convince ourselves of the absence of complete laws of behaviour as of complete rules of conduct. The only way we know of for finding such laws is scientific observation, and we certainly know of no circumstances under which we could say, ' We have searched enough. There are no such laws.'

We can demonstrate more forcibly that any such statement would be unjustified. For suppose we could be sure of finding such laws if they existed. Then given a discrete-state machine it should certainly be possible to discover by observation sufficent about it to predict its future behaviour, and this within a reasonable time, say a thousand years. But this does not seem to be the case. I have set up on the Manchester computer a small programme using only 1000 units of storage, whereby the machine supplied with one sixteen figure number replies with another within two seconds. I would defy anyone to learn from these replies sufficient about the programme to be able to predict any replies to untried values.

(9) *The Argument from Extra-Sensory Perception.* I assume that the reader is familiar with the idea of extra-sensory perception, and the meaning of the four items of it, *viz.* telepathy, clairvoyance, precognition and psycho-kinesis. These disturbing phenomena seem to deny all our usual scientific ideas. How we should like to discredit them ! Unfortunately the statistical evidence, at least for telepathy, is overwhelming. It is very difficult to rearrange one's ideas so as to fit these new facts in. Once one has accepted them it does not seem a very big step to believe in ghosts and bogies. The idea that our bodies move simply according to the known laws of physics, together with some others not yet discovered but somewhat similar, would be one of the first to go.

This argument is to my mind quite a strong one. One can say in reply that many scientific theories seem to remain workable in practice, in spite of clashing with E.S.P. ; that in fact one can get along very nicely if one forgets about it. This is rather cold comfort, and one fears that thinking is just the kind of phenomenon where E.S.P. may be especially relevant.

A more specific argument based on E.S.P. might run as follows : " Let us play the imitation game, using as witnesses a man who is good as a telepathic receiver, and a digital computer. The interrogator can ask such questions as ' What suit does the card in my right hand belong to ? ' The man by telepathy or clairvoyance gives the right answer 130 times out of 400 cards. The machine can only guess at random, and perhaps gets 104 right, so the interrogator makes the right identification." There is an interesting possibility which opens here. Suppose the digital computer contains a random number generator. Then it will be natural to use this to decide what answer to give. But then the random number generator will be subject to the psycho-kinetic powers of the interrogator. Perhaps this psycho-kinesis might cause the machine to guess right more often than would be expected on a probability calculation, so that the interrogator might still be unable to make the right identification. On the other hand, he might be able to guess right without any questioning, by clairvoyance. With E.S.P. anything may happen.

If telepathy is admitted it will be necessary to tighten our test up. The situation could be regarded as analogous to that which would occur if the interrogator were talking to himself and one of the competitors was listening with his ear to the wall. To put the competitors into a ' telepathy-proof room ' would satisfy all requirements.

7. *Learning Machines.*

The reader will have anticipated that I have no very convincing arguments of a positive nature to support my views. If I had I should not have taken such pains to point out the fallacies in contrary views. Such evidence as I have I shall now give.

Let us return for a moment to Lady Lovelace's objection, which stated that the machine can only do what we tell it to do. One could say that a man can ' inject ' an idea into the machine, and that it will respond to a certain extent and then drop into quiescence, like a piano string struck by a hammer. Another simile would be an atomic pile of less than critical size : an injected idea is to correspond to a neutron entering the pile from without. Each such neutron will cause a certain disturbance which eventually dies away. If, however, the size of the pile is sufficiently increased, the disturbance caused by such an incoming neutron will very likely go on and on increasing until the whole pile is destroyed. Is there a corresponding phenomenon for minds, and is there one for machines ? There does seem to be one for the human mind. The majority of them seem to be ' sub-critical ', *i.e.* to correspond in this analogy to piles of sub-critical size. An idea presented to such a mind will on average give rise to less than one idea in reply. A smallish proportion are super-critical. An idea presented to such a mind may give rise to a whole ' theory ' consisting of secondary, tertiary and more remote ideas. Animals minds seem to be very definitely sub-critical. Adhering to this analogy we ask, ' Can a machine be made to be super-critical ?'

The ' skin of an onion ' analogy is also helpful. In considering the functions of the mind or the brain we find certain operations which we can explain in purely mechanical terms. This we say does not correspond to the real mind : it is a sort of skin which we must strip off if we are to find the real mind. But then in what remains we find a further skin to be stripped off, and so on.

Proceeding in this way do we ever come to the 'real' mind, or do we eventually come to the skin which has nothing in it ? In the latter case the whole mind is mechanical. (It would not be a discrete-state machine however. We have discussed this.)

These last two paragraphs do not claim to be convincing arguments. They should rather be described as 'recitations tending to produce belief'.

The only really satisfactory support that can be given for the view expressed at the beginning of § 6, will be that provided by waiting for the end of the century and then doing the experiment described. But what can we say in the meantime ? What steps should be taken now if the experiment is to be successful ?

As I have explained, the problem is mainly one of programming. Advances in engineering will have to be made too, but it seems unlikely that these will not be adequate for the requirements. Estimates of the storage capacity of the brain vary from 10^{10} to 10^{15} binary digits. I incline to the lower values and believe that only a very small fraction is used for the higher types of thinking. Most of it is probably used for the retention of visual impressions. I should be surprised if more than 10^9 was required for satisfactory playing of the imitation game, at any rate against a blind man. (Note—The capacity of the *Encyclopaedia Britannica*, 11th edition, is 2×10^9.) A storage capacity of 10^7 would be a very practicable possibility even by present techniques. It is probably not necessary to increase the speed of operations of the machines at all. Parts of modern machines which can be regarded as analogues of nerve cells work about a thousand times faster than the latter. This should provide a 'margin of safety' which could cover losses of speed arising in many ways. Our problem then is to find out how to programme these machines to play the game. At my present rate of working I produce about a thousand digits of programme a day, so that about sixty workers, working steadily through the fifty years might accomplish the job, if nothing went into the waste-paper basket. Some more expeditious method seems desirable.

In the process of trying to imitate an adult human mind we are bound to think a good deal about the process which has brought it to the state that it is in. We may notice three components,

(a) The initial state of the mind, say at birth,

(b) The education to which it has been subjected,

(c) Other experience, not to be described as education, to which it has been subjected.

[[13]]

Instead of trying to produce a programme to simulate the adult mind, why not rather try to produce one which simulates the child's ? If this were then subjected to an appropriate course of education one would obtain the adult brain. Presumably the child-brain is something like a note-book as one buys it from the stationers. Rather little mechanism, and lots of blank sheets. (Mechanism and writing are from our point of view almost synonymous.) Our hope is that there is so little mechanism in the child-brain that something like it can be easily programmed. The amount of work in the education we can assume, as a first approximation, to be much the same as for the human child.

We have thus divided our problem into two parts. The child-programme and the education process. These two remain very closely connected. We cannot expect to find a good child-machine at the first attempt. One must experiment with teaching one such machine and see how well it learns. One can then try another and *see* if it is better or worse. There is an obvious connection between this process and evolution, by the identifications

Structure of the child machine = Hereditary material
Changes   ,,   ,,   = Mutations
Natural selection   = Judgment of the experimenter

One may hope, however, that this process will be more expeditious than evolution. The survival of the fittest is a slow method for measuring advantages. The experimenter, by the exercise of intelligence, should be able to speed it up. Equally important is the fact that he is not restricted to random mutations. If he can trace a cause for some weakness he can probably think of the kind of mutation which will improve it.

It will not be possible to apply exactly the same teaching process to the machine as to a normal child. It will not, for instance, be provided with legs, so that it could not be asked to go out and fill the coal scuttle. Possibly it might not have eyes. But however well these deficiencies might be overcome by clever engineering, one could not send the creature to school without the other children making excessive fun of it. It must be given some tuition. We need not be too concerned about the legs, eyes, etc. The example of Miss *Helen Keller* shows that education can take place provided that communication in both directions between teacher and pupil can take place by some means or other.

[[14]]

We normally associate punishments and rewards with the teaching process. Some simple child-machines can be constructed or programmed on this sort of principle. The machine has to be so constructed that events which shortly preceded the occurrence of a punishment-signal are unlikely to be repeated, whereas a reward-signal increased the probability of repetition of the events which led up to it. These definitions do not presuppose any feelings on the part of the machine. I have done some experiments with one such child-machine, and succeeded in teaching it a few things, but the teaching method was too unorthodox for the experiment to be considered really successful.

The use of punishments and rewards can at best be a part of the teaching process. Roughly speaking, if the teacher has no other means of communicating to the pupil, the amount of information which can reach him does not exceed the total number of rewards and punishments applied. By the time a child has learnt to repeat 'Casabianca' he would probably feel very sore indeed, if the text could only be discovered by a 'Twenty Questions' technique, every 'NO' taking the form of a blow. It is necessary therefore to have some other 'unemotional' channels of communication. If these are available it is possible to teach a machine by punishments and rewards to obey orders given in some language, *e.g.* a symbolic language. These orders are to be transmitted through the 'unemotional' channels. The use of this language will diminish greatly the number of punishments and rewards required.

Opinions may vary as to the complexity which is suitable in the child machine. One might try to make it as simple as possible consistently with the general principles. Alternatively one might have a complete system of logical inference 'built in'.[1] In the latter case the store would be largely occupied with definitions and propositions. The propositions would have various kinds of status, *e.g.* well-established facts, conjectures, mathematically proved theorems, statements given by an authority, expressions having the logical form of proposition but not belief-value. Certain propositions may be described as 'imperatives'. The machine should be so constructed that as soon as an imperative is classed as 'well-established' the appropriate action automatically takes place. To illustrate this, suppose the teacher says to the machine, 'Do your homework now'. This may cause "Teacher says 'Do your homework now'" to be included amongst the well-established facts. Another such fact might be,

---

[1] Or rather 'programmed in' for our child-machine will be programmed in a digital computer. But the logical system will not have to be learnt.

" Everything that teacher says is true ". Combining these may
eventually lead to the imperative, ' Do your homework now ',
being included amongst the well-established facts, and this,
by the construction of the machine, will mean that the homework
actually gets started, but the effect is very satisfactory. The
processes of inference used by the machine need not be such
as would satisfy the most exacting logicians. There might for
instance be no hierarchy of types. But this need not mean that
type fallacies will occur, any more than we are bound to fall
over unfenced cliffs. Suitable imperatives (expressed *within*
the systems, not forming part of the rules *of* the system) such
as ' Do not use a class unless it is a subclass of one which has
been mentioned by teacher ' can have a similar effect to ' Do not
go too near the edge '.

The imperatives that can be obeyed by a machine that has
no limbs are bound to be of a rather intellectual character, as in
the example (doing homework) given above. Important amongst
such imperatives will be ones which regulate the order in which
the rules of the logical system concerned are to be applied.
For at each stage when one is using a logical system, there is a
very large number of alternative steps, any of which one is
permitted to apply, so far as obedience to the rules of the logical
system is concerned. These choices make the difference between
a brilliant and a footling reasoner, not the difference between a
sound and a fallacious one. Propositions leading to imperatives
of this kind might be " When Socrates is mentioned, use the
syllogism in Barbara " or " If one method has been proved to be
quicker than another, do not use the slower method ". Some
of these may be ' given by authority ', but others may be produced by the machine itself, *e.g.* by scientific induction.

The idea of a learning machine may appear paradoxical to
some readers. How can the rules of operation of the machine
change ? They should describe completely how the machine
will react whatever its history might be, whatever changes
it might undergo. The rules are thus quite time-invariant.
This is quite true. The explanation of the paradox is that the
rules which get changed in the learning process are of a rather
less pretentious kind, claiming only an ephemeral validity. The
reader may draw a parallel with the Constitution of the United
States.

An important feature of a learning machine is that its teacher
will often be very largely ignorant of quite what is going on
inside, although he may still be able to some extent to predict
his pupil's behaviour. This should apply most strongly to the

later education of a machine arising from a child-machine of well-tried design (or programme). This is in clear contrast with normal procedure when using a machine to do computations : one's object is then to have a clear mental picture of the state of the machine at each moment in the computation. This object can only be achieved with a struggle. The view that 'the machine can only do what we know how to order it to do ',[1] appears strange in face of this. Most of the programmes which we can put into the machine will result in its doing something that we cannot make sense of at all, or which we regard as completely random behaviour. Intelligent behaviour presumably consists in a departure from the completely disciplined behaviour involved in computation, but a rather slight one, which does not give rise to random behaviour, or to pointless repetitive loops. Another important result of preparing our machine for its part in the imitation game by a process of teaching and learning is that ' human fallibility ' is likely to be omitted in a rather natural way, *i.e.* without special 'coaching'. (The reader should reconcile this with the point of view on pp. 24, 25.) Processes that are learnt do not produce a hundred per cent. certainty of result ; if they did they could not be unlearnt.

It is probably wise to include a random element in a learning machine (see p. 438). A random element is rather useful when we are searching for a solution of some problem. Suppose for instance we wanted to find a number between 50 and 200 which was equal to the square of the sum of its digits, we might start at 51 then try 52 and go on until we got a number that worked. Alternatively we might choose numbers at random until we got a a good one. This method has the advantage that it is unnecessary to keep track of the values that have been tried, but the disadvantage that one may try the same one twice, but this is not very important if there are several solutions. The systematic method has the disadvantage that there may be an enormous block without any solutions in the region which has to be investigated first. Now the learning process may be regarded as a search for a form of behaviour which will satisfy the teacher (or some other criterion). Since there is probably a very large number of satisfactory solutions the random method seems to be better than the systematic. It should be noticed that it is used in the analogous process of evolution. But there the systematic method is not possible. How could one keep track

---

[1] Compare Lady Lovelace's statement (p. 450), which does not contain the word ' only '.

of the different genetical combinations that had been tried, so as to avoid trying them again ?

We may hope that machines will eventually compete with men in all purely intellectual fields. But which are the best ones to start with ? Even this is a difficult decision. Many people think that a very abstract activity, like the playing of chess, would be best. It can also be maintained that it is best to provide the machine with the best sense organs that money can buy, and then teach it to understand and speak English. This process could follow the normal teaching of a child. Things would be pointed out and named, etc. Again I do not know what the right answer is, but I think both approaches should be tried.

We can only see a short distance ahead, but we can see plenty there that needs to be done.

BIBLIOGRAPHY

Samuel Butler, Erewhon, London, 1865. Chapters 23, 24, 25, *The Book of the Machines*.

Alonzo Church, " An Unsolvable Problem of Elementary Number Theory ", *American J. of Math.*, 58 (1936), 345-363.

K. Gödel, " Über formal unentscheidbare Sätze der Principia Mathematica und verwandter Systeme, I ", *Monatshefte für Math. und Phys.*, (1931), 173-189.

D. R. Hartree, *Calculating Instruments and Machines*, New York, 1949.

S. C. Kleene, " General Recursive Functions of Natural Numbers ", *American J. of Math.*, 57 (1935), 153-173 and 219-244.

G. Jefferson, " The Mind of Mechanical Man ". Lister Oration for 1949. *British Medical Journal*, vol. i (1949), 1105-1121.

Countess of Lovelace, ' Translator's notes to an article on Babbage's Analytical Engine ', *Scientific Memoirs* (ed. by R. Taylor), vol. 3 (1842), 691-731.

Bertrand Russell, *History of Western Philosophy*, London, 1940.

A. M. Turing, " On Computable Numbers, with an Application to the Entscheidungsproblem ", *Proc. London Math. Soc.* (2), 42 (1937), 230-265.

Victoria University of Manchester.

Chapter 25

DIGITAL COMPUTERS APPLIED
TO GAMES

Chess problems are the hymn tunes of mathematics—G. H. HARDY

MACHINES WHICH WILL PLAY GAMES have a long and interesting history. Among the first and most famous was the chess-playing automaton constructed in 1769 by the Baron Kempelen; M. Maelzel took it on tour all over the world, deceiving thousands of people into thinking that it played the game automatically. This machine was described in detail by Edgar Allan Poe; it is said to have defeated Napoleon himself—and he was accounted quite a good player, but it was finally shown up when somebody shouted "FIRE" during a game, and caused the machine to go into a paroxysm owing to the efforts of the little man inside to escape.

In about 1890 Signor Torres Quevedo made a simple machine—a real machine this time—which with a rook and king can checkmate an opponent with a single king. This machine avoids stalemate very cleverly and always wins its games. It allows an opponent to make two mistakes before it refuses to play further with him, so it is always possible to cheat by moving one's own king the length of the board. The mechanism of the machine is such that it cannot move its rook back past its king and one can then force a draw! This machine, like Babbage's "noughts and crosses" machine is relatively simple, the rules to be obeyed are quite straightforward, and the machines couldn't lose. Babbage thought that his analytical engine ought to be able to play a real game of chess, which is a much more difficult thing to do.

In this chapter we discuss how a digital computer can be made to play chess—it does so rather badly, and how it plays draughts— it does so quite well. We shall also describe a special simple machine which was built to entertain the public during the Festival of Britain. It was called Nimrod because it played nim, a game which is like noughts and crosses, in that the tricks which are needed to win can be expressed in mathematical terms. This machine was on show in South Kensington for six months and took on all comers.

286

During the Festival the Society for Psychical Research came and fitted up a room nearby in order to see if the operations of the machine could be influenced by concentrated thought on the part of the research workers, most of whom were elderly ladies. When this experiment had failed they tried to discover whether they in turn could be affected by vibrations from the machine, and could tell from another room how the game was progressing. Unfortunately this experiment, like the first, was a complete failure, the only conclusion being that machines are much less co-operative than human beings in telepathic experiments.

At the end of the Festival of Britain Nimrod was flown to Berlin and shown at the Trade Fair. The Germans had never seen anything like it, and came to see it in their thousands, so much so in fact that on the first day of the show they entirely ignored a bar at the far end of the room where free drinks were available, and it was necessary to call out special police to control the crowds. The machine became even more popular after it had defeated the Economics Minister, Dr. Erhardt, in three straight games. After this it was taken to Canada and demonstrated to the Society of Engineers in Toronto. It is still somewhere on the North American continent, though it may have been dismantled by now, and it would be amusing to match it against some of the other nim-playing machines which have been built in the last year or two.

The reader might well ask why we bother to use these complicated and expensive machines in so trivial a pursuit as playing games. It would be disingenuous of us to disguise the fact that the principal motive which prompted the work was the sheer fun of the thing, but nevertheless if ever we had to justify the time and effort (and we feel strongly that no excuses are either necessary or called for) we could quite easily make a pretence at doing so. We have already explained how hard all programming is to do, and how much difficulty is due to the incompetence of the machine at taking an overall view of the problem which it is analysing. This particular point is brought out more clearly in playing games than in anything [1] else. The machine cannot look at the whole of a chess board at once; it has to peer short-sightedly at every square in turn, in much the same way as it has to look at a commercial document. Research into the techniques of programming a machine to tackle complicated problems of this type may in fact lead to quite important [2] advances, and help in serious work in business and economics— perhaps, regrettably, even in the theory of war. We hope that mathematicians will continue to play draughts and chess, and to enjoy themselves as long as they can.

We have often been asked why we don't use the machine to work out the football pools, or even to do something to remove the present uncertainty about the results of tomorrow's horse races. ⟦3⟧ Perhaps one day we shall persuade our mathematicians to apply themselves to this problem too. It would first be necessary to establish a series of numerical criteria from which the machine could predict the results with greater certainty than the ordinary citizen can achieve with his pin; the presumption underlying the whole idea is that such criteria do in fact exist, but that they are too complicated for a man to apply in time, whereas a machine could do the necessary computations for him. It is most unlikely that a machine could ever hope to predict (for example) the results of a single football match, but it is at least possible that a detailed analysis might significantly improve the odds in favour of the gambler, so that if he invested on a large enough scale he could make a profit. It is notoriously true that mathematics, and particularly the theory of probability, owes more to gambling than gambling owes to mathematics; perhaps a machine might do something to restore the balance. Lady Lovelace lost a fortune by trying to back horses scientifically, and many others have done the same; all one could hope for is a slight improvement in the odds. We might make it pay but we doubt it; as an academic exercise it would be amusing, but we shall give the project a low priority.

CHESS

When one is asked, "Could one make a machine to play chess?" there are several possible meanings which might be given to the words. Here are a few—

(*a*) Could one make a machine which would obey the rules of ⟦4⟧ chess, i.e. one which would play random legal moves, or which could tell one whether a given move is a legal one?

(*b*) Could one make a machine which would solve chess prob- ⟦5⟧ lems, e.g. tell one whether, in a given position, white has a forced mate in three?

(*c*) Could one make a machine which would play a reasonably ⟦6⟧ good game of chess, i.e. which, confronted with an ordinary (that is, not particularly unusual) chess position, would after two or three minutes of calculation, indicate a passably good legal move?

[[7]]   (*d*) Could one make a machine to play chess, and to improve its play, game by game, profiting from its experience?

To these we may add two further questions, unconnected with chess, which are likely to be on the tip of the reader's tongue.

[[8]]   (*e*) Could one make a machine which would answer questions put to it, in such a way that it would not be possible to distinguish its answers from those of a man?

[[9]]   (*f*) Could one make a machine which would have feelings as you and I have?

The problem to be considered here is (*c*), but to put this problem into perspective with the others I shall give the very briefest of answers to each of them.

To (*a*) and (*b*) I should say, "This certainly can be done. If it has not been done already it is merely because there is something better to do."

Question (*c*) we are to consider in greater detail, but the short answer is, "Yes, but the better the standard of play required, the more complex will the machine be, and the more ingenious perhaps the designer."

To (*d*) and (*e*) I should answer, "I believe so. I know of no really convincing argument to support this belief, and certainly of none to disprove it."

To (*f*) I should say, "I shall never know, any more than I shall ever be quite certain that *you* feel as I do."

In each of these problems except possibly the last, the phrase, "Could one make a machine to . . ." might equally well be replaced by, "Could one programme an electronic computer to . . ." Clearly the electronic computer so programmed would itself constitute a machine. And on the other hand if some other machine had been constructed to do the job we could use an electronic computer (of sufficient storage capacity), suitably programmed, to calculate what this machine would do, and in particular what answer it would give.

After these preliminaries let us give our minds to the problem of making a machine, or of programming a computer, to play a tolerable game of chess. In this short discussion it is of course out of the question to provide actual programmes, but this does not really matter on account of the following principle—

If one can explain quite unambiguously in English, with the aid of mathematical symbols if required, how a calculation is to be done, then it is always possible to programme any digital computer to do that calculation, provided the storage capacity is adequate.

This is not the sort of thing that admits of clear-cut proof, but amongst workers in the field it is regarded as being clear as day. Accepting this principle, our problem is reduced to explaining "unambiguously in English" the rules by which the machine is to choose its move in each position. For definiteness we will suppose the machine is playing white.

If the machine could calculate at an infinite speed, and also had unlimited storage capacity, a comparatively simple rule would suffice, and would give a result that in a sense could not be improved on. This rule could be stated:

"Consider every possible continuation of the game from the given position. There is only a finite number of them (at any rate if the fifty-move rule makes a draw obligatory, not merely permissive). Work back from the end of these continuations, marking a position with white to play as 'win' if there is a move which turns it into a position previously marked as 'win.' If this does not occur, but there is a move which leads to a position marked 'draw,' then mark the position 'draw.' Failing this, mark it 'lose.' Mark a position with black to play by a similar rule with 'win' and 'lose' interchanged. If after this process has been completed it is found that there are moves which lead to a position marked 'win,' one of these should be chosen. If there is none marked 'win' choose one marked 'draw' if such exists. If all moves lead to a position marked 'lose,' any move may be chosen."

Such a rule is practically applicable in the game of noughts and crosses, but in chess is of merely academic interest. Even when the rule can be applied it is not very appropriate for use against a weak opponent, who may make mistakes which ought to be exploited.   [10]

In spite of the impracticability of this rule it bears some resemblance to what one really does when playing chess. One does not follow all the continuations of play, but one follows some of them. One does not follow them until the end of the game, but one follows them a move or two, perhaps more. Eventually a position seems, rightly or wrongly, too bad to be worth further consideration, or (less frequently) too good to hesitate longer over. The further a position is from the one on the board the less likely it is to occur, and therefore the shorter is the time which can be assigned for its consideration. Following this idea we might have a rule something like this—

"Consider all continuations of the game consisting of a move by white, a reply by black, and another move and reply. The value of the position at the end of each of these sequences of moves is estimated according to some suitable rule. The values at earlier positions are then calculated by working backwards move by move as in the theoretical rule given before. The move to be chosen is that which leads to the position with the greatest value."

[[11]]

It is possible to arrange that no two positions have the same value. The rule is then unambiguous. A very simple form of values, but one not having this property, is an "evaluation of material," e.g. on the basis—

$$P = 1$$
$$Kt = 3$$
$$B = 3\tfrac{1}{2}$$
$$R = 5$$
$$Q = 10$$
$$\text{Checkmate} = 1000$$

If B is black's total and W is white's, then W/B is quite a good measure of value. This is better than $W - B$ as the latter does not encourage exchanges when one has the advantage. Some small extra arbitrary function of position may be added to ensure definiteness in the result.

The weakness of this rule is that it follows all combinations equally far. It would be much better if the more profitable moves were considered in greater detail than the less. It would also be desirable to take into account more than mere "value of material."

After this introduction I shall describe a particular set of rules, which could without difficulty be made into a machine programme. It is understood that the machine is white and that white is next to play. The current position is called the *position on the board*, and the positions arising from it by later moves *positions in the analysis*.

"CONSIDERABLE" MOVES

"Considerable" here is taken to mean moves which will be "considered" in the analysis by the machine.

Every possibility for white's next move and for black's reply is "considerable." If a capture is considerable then any recapture is considerable. The capture of an undefended piece or the capture of a piece of higher value by one of lower value is always considerable. A move giving checkmate is considerable.

DEAD POSITION

A position in the analysis is dead if there are no considerable moves in that position, i.e. if it is more than two moves ahead of the

present position, and no capture or recapture or mate can be made   ⟦12⟧
in the next move.

VALUE OF POSITION

The value of a dead position is obtained by adding up the piece values as above, and forming the ratio W/B of white's total to black's. In other positions with white to play the value is the greatest value of (*a*) the positions obtained by considerable moves, or (*b*) the position itself evaluated as if a dead position. The latter alternative is to be omitted if all moves are considerable. The same process is to be undertaken for one of black's moves, but the machine will then choose the *least* value.

POSITION-PLAY VALUE

Each white piece has a certain position-play contribution and so has the black king. These must all be added up to give the position-play value.

For a Q, R, B, or Kt, count—

(*a*) The square root of the number of moves the piece can make from the position, counting a capture as two moves, and not forgetting that the king must not be left in check.

(*b*) (If not a Q) 1·0 if it is defended, and an additional 0·5 if twice defended.

For a K, count—

(*c*) For moves other than castling as (*a*) above.

(*d*) It is then necessary to make some allowance for the vulnerability of the K. This can be done by assuming it to be replaced by a friendly Q on the same square, estimating as in (*a*), but subtracting instead of adding.

(*e*) Count 1·0 for the possibility of castling later not being lost by moves of K or rooks, a further 1·0 if castling could take place on the next move, and yet another 1·0 for the actual performance of castling.

For a P, count—

(*f*) 0·2 for each rank advanced.

(*g*) 0·3 for being defended by at least one piece (not P).

For the black K, count—

(*h*) 1·0 for the threat of checkmate.

(*i*) 0·5 for check.   ⟦13⟧

We can now state the rule for play as follows. The move chosen must have the greatest possible value, and, consistent with this, the greatest possible position-play value. If this condition admits of several solutions a choice may be made at random, or according to an arbitrary additional condition.

Note that no "analysis" is involved in position-play evaluation. This is to reduce the amount of work done on deciding the move.

The game below was played between this machine and a weak player who did not know the system. To simplify the calculations the square roots were rounded off to one decimal place, i.e. this table was used—

Number . . .	0	1	2	3	4	5	6	7	8	9	10	11	12	13
Square Root . .	0	1	1·4	1·7	2·0	2·2	2·4	2·6	2·8	3·0	3·2	3·3	3·5	3·6

No random choices actually arose in this game. The increase of position-play value is given after white's move if relevant. An asterisk indicates that every other move had a lower position-play value.

	White (Machine)		Black
1.	P—K 4	4·2*	P—K 4
2.	Kt—Q B 3	3·1*	Kt—K B 3
3.	P—Q 4	2·6*	B—Q Kt 5
4.	Kt—K B 3[1]	2·0	P—Q 3
5.	B—Q 2	3·5*	Kt—Q B 3
6.	P—Q 5	0·2	Kt—Q 5
7.	P—K R 4[2]	1·1*	B—Kt 5
8.	P—Q R 4[2]	1·0*	Kt × Kt ch.
9.	P × Kt		B—K R 4
10.	B—Kt 5 ch.	2·4*	P—Q B 3
11.	P × P		O—O
12.	P × P		R—Kt 1
13.	B—R 6	− 1·5	Q—R 4
14.	Q—K 2	0·6	Kt—Q 2
15.	K R—Kt 1[3]	1·2*	Kt—B 4[4]
16.	R—Kt 5[5]		B—Kt 3
17.	B—Kt 5	0·4	Kt × Kt P
18.	O—O—O	3·2*	Kt—B 4
19.	B—B 6		K R—Q B 1
20.	B—Q 5		B × Kt
21.	B × B	0·7	Q × P
22.	K—Q 2		Kt—K 3
23.	R—Kt 4	−0·3	Kt—Q 5
24.	Q—Q 3		Kt—Kt 4
25.	B—Kt 3		Q—R 3
26.	B—B 4		B—R 4
27.	R—Kt 3		Q—R 5
28.	B × Kt		Q × B
29.	Q × P [6]		R—Q 1[4]
30.	Resigns[7]		

Notes—

1. If B—Q 2 3·7* then P × P is foreseen.
2. Most inappropriate moves.
3. If white castles then B × Kt, B × B, Q × P.
4. The fork is unforeseen at white's last move.
5. Heads in the sand!
6. Fiddling while Rome burns!
7. On the advice of his trainer.

Numerous criticisms of the machine's play may be made. It is quite defenceless against forks, although it may be able to see certain other kinds of combination. It is of course not difficult to devise improvements of the programme so that these simple forks are foreseen. The reader may be able to think of some such improvements for himself. Since no claim is made that the above rule is particularly good, I have been content to leave this flaw without remedy; clearly a line has to be drawn between the flaws which one will attempt to eliminate and those which must be accepted as a risk. Another criticism is that the scheme proposed, although reasonable in the middle game, is futile in the end game. The change-over from the middle game to the end-game is usually sufficiently clear-cut for it to be possible to have an entirely different system for the end-game. [[14]] This should of course include quite definite programmes for the standard situations, such as mate with rook and king, or king and pawn against king. There is no intention to discuss the end-game further here.

If I were to sum up the weakness of the above system in a few words I would describe it as a caricature of my own play. It was in fact based on an introspective analysis of my thought processes when playing, with considerable simplifications. It makes oversights which are very similar to those which I make myself, and which may in both cases be ascribed to the considerable moves being inappropriately chosen. This fact might be regarded as supporting the glib view which is often expressed, to the effect that "one cannot programme a machine to play a better game than one plays oneself." This statement should I think be compared with another [[15]] of rather similar form. "No animal can swallow an animal heavier than himself." Both statements are, as far as I know, untrue. They are also both of a kind that one is easily bluffed into accepting, partly because one thinks that there ought to be some slick way of demonstrating them, and one does not like to admit that one does not see what this argument is. They are also both supported by normal experience, and need exceptional cases to falsify them. The statement about chess programming may be falsified quite simply by the speed of the machine, which might make it feasible to carry the analysis a move farther than a man could do in the same time. This effect is less than might be supposed. Although electronic computers are very fast where conventional computing is concerned, their advantage is much reduced where enumeration of cases, etc., is involved on a large scale. Take for instance the problem of counting the possible moves from a given position in chess. If the number is 30 a man might do it in 45 seconds and the machine in 1 second. The machine has still an advantage, but it is much less overwhelming than it would be for instance when calculating cosines.

In connexion with the question of the ability of a chess-machine to profit from experience, one can see that it would be quite possible to programme the machine to try out variations in its method of play (e.g. variations in piece value) and adopt the one ⟦16⟧ giving the most satisfactory results. This could certainly be described as "learning," though it is not quite representative of learning as we know it. It might also be possible to programme the machine to search for new types of combination in chess. If this project produced results which were quite new, and also interesting to the programmer, who should have the credit? Compare this with the situation where a Defence Minister gives orders for research to be done to find a counter to the bow and arrow. Should the inventor of the shield have the credit, or should the Defence Minister?

THE MANCHESTER UNIVERSITY MACHINE

In November, 1951, some months after this article was written (by Dr. Turing) Dr. Prinz was able to make the Manchester University machine solve a few straightforward chess problems of the "Mate-in-Two" type (see *Research*, Vol. 6 (1952), p. 261).

It is usually true to say that the best and often the only way to see how well the machine can tackle a particular type of problem is to produce a definite programme for the machine, and, in this case, in order to have something working in the shortest possible time, a few restrictions were imposed on the rules of chess as they were "explained" to the machine. For example castling was not permitted, nor were double moves by pawns, nor taking *en passant* nor the promotion of a pawn into a piece when it reached the last row; further, no distinction was made between mate and stalemate.

⟦17⟧ The programme contained a routine for the construction of the next possible move, a routine to check this move for legality, and various sequences for recording the moves and the positions obtained. All these separate subroutines were linked together by a master routine which reflected the structure of the problem as a whole and ensured that the subroutines were entered in the proper sequence.

The technique of programming was perhaps rather crude, and many refinements, increasing the speed of operation, are doubtless possible. For this reason, the results reported here can only serve as a very rough guide to the speed attainable; but they do show the need for considerable improvement in programming technique and machine performance before a successful game by a machine against a human chess player becomes a practical possibility.

The programme, as well as the initial position on the chess board, was supplied to the machine on punched tape and then transferred to the magnetic store of the machine.

A initial routine (sub-programme) was transferred to the electronic store, and the machine started its computation. The programme was so organized that every first move by white was printed out; after the key move had been reached the machine printed: "MATE."

The main result of the experiment was that the machine is disappointingly slow when playing chess—in contrast to the extreme superiority over human computers where purely mathematical problems are concerned. For the simple example given in the position reproduced here, 15 minutes were needed to print the solution. A detailed analysis shows that the machine tried about 450 possible moves (of which about 100 were illegal) in the course of the game; this means about two seconds per move on the average.

A considerable portion of this time had to be used for a test for self-check (i.e. after a player had made a move, to find out whether his own King was left in check). This was done by first examining all squares connected to the King's square by a Knight's move, to see (a) whether they were on the board at all, (b) whether they were empty or occupied, (c) if occupied, by a piece of which colour and (d) if occupied by a piece of opposite colour, whether or not this piece was a Knight. A similar test had to be carried out for any other piece that might have put the King in check. This test involves several hundreds of operations and, at a machine speed of 1 msec per operation, might take an appreciable fraction of a second.

The next important time-consuming factor was the magnetic transfers, i.e. the transfers of sub-programmes and data (relating to positions and moves) between the magnetic and the electronic store. It is here that improved programming technique may save time by better utilization of the electronic store, thus reducing the number of transfers (nine for every legal move in the present programme). ⟦18⟧

Compared with these two items, the time spent in computing the moves appeared to be of minor importance although the machine not only computed the possible moves but also the impossible, but "thinkable" moves—meaning those which either carry the piece off the board, or lead to a collision with a piece of the same colour already on the square. These moves, however, were quickly rejected by the machine and did not contribute greatly to the total computation time.

It appears that if this crude method of programming were the only one available it would be quite impractical for any machine to compete on reasonable terms with a competent human being.

Before we conclude too easily that no computer will ever compete in a Masters' Tournament let us remind ourselves that the Manchester machine solved a problem after a few weeks tuition, which represents quite reasonable progress for a beginner.

The First Chess Problem Solved by a Computing Machine. The task set the Manchester machine was to find a move by white that would lead to a mate in the next move, whatever black might answer. The move is R—R6.

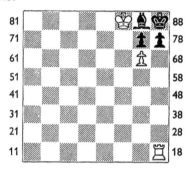

For solution of the problem by the machine the squares of the board were numbered in rather unusual fashion. The bottom row was numbered 11 to 18 (from left to right), the next 21 to 28, and so on to the top row, which was 81–88. Square 68 was thus the square in row 6, column 8. The machine has printed out all the moves which white tried out to find a solution, and has printed "MATE" after finding and recording the key move, which appears in the form "Rook to 68."

The list of moves is—

Pawn to 78.	Rook to 11.
Rook to 17.	Rook to 28.
Rook to 16.	Rook to 38.
Rook to 15.	Rook to 48.
Rook to 14.	Rook to 58.
Rook to 13.	Rook to 68.
Rook to 12.	MATE.

The game of draughts occupies an intermediate position between   [[19]]
the extremely complex games such as chess, and the relatively
simple games such as nim or noughts-and-crosses for which a complete mathematical theory exists. This fact makes it a rather suitable
subject for experiments in mechanical game playing, for although
there is no complete theory of the game available, so that the machine
has to look ahead to find the moves, the moves themselves are rather
simple and relatively few in number.

Various forms of strategy have been suggested for constructing
an automatic chess player; the purpose of such plans is to reduce the
time taken by the machine to choose its move. As Prinz has shown,
the time taken by any machine which considers all the possible moves
for four or five steps ahead would be quite prohibitive, and the
principal aim of the strategy is to reduce this number very considerably, while at the same time introducing a scheme of valuing the
positions which will allow it to choose a reasonably good move. The
chief interest in games-playing machines lies in the development of a
suitable strategy.

Before any strategy can be realized in practice, however, the
basic programme necessary to find the possible moves and to make
them must be constructed. When this has been done the strategy,
which consists principally of the methods by which positions can be
valued, can be added to make the complete game player. It is
obviously possible to make experiments with different strategies
using the same basic move-finding-and-making routine.

The basic programme for draughts, which is described in outline
in the following paragraphs, is very much simpler than the corresponding one for chess. It has in fact proved possible to put both it
and the necessary position storage in the electronic store of the
Manchester machine at the same time. This removes the need for
magnetic transfers during the operation of the programme, and
this fact, together with the simplicity of the moves, has reduced
the time taken to consider a single move to about one tenth of a
second.

BASIC PROGRAMME FOR DRAUGHTS

We must first consider the representation of a position in the
machine. The 32 squares used in a draughts board are numbered as
shown in the diagram.

A position is represented by 3 thirty-two-digit binary numbers (or words) B, W and K which give the positions of the black men (and kings), the white men (and kings) and the kings (of either colour) respectively. The digits of these words each represent a square on the board; the square n being represented by the digit 2^n.

BLACK

0	1	2	3
4	5	6	7
8	9	10	11
12	13	14	15
16	17	18	19
20	21	22	23
24	25	26	27
28	29	30	31

WHITE

Thus the least significant digit represents square 0 and the most significant digit represents square 31. (In the Manchester machine, where the word length is 40 digits, the last 8 digits are irrelevant). A unit in the word indicates the presence, and a zero indicates the absence of the appropriate type of man in the corresponding square. Thus the opening position of the game would be represented by*—

$$B = 1111, 1111, 1111, 0000, 0000, 0000, 0000, 0000$$

$$W = 0000, 0000, 0000, 0000, 0000, 1111, 1111, 1111$$

$$K = 0000, 0000, 0000, 0000, 0000, 0000, 0000, 0000$$

The positions of the white kings are indicated by the word $W\&K$, while the empty squares are indicated by the word $\sim W\&\sim B$.†

It will be seen that there are at most four possible types of non-capture moves from any square on the board. For example, from square 14 the possible moves are to squares 9, 10, 17 or 18. The machine considers all these moves in turn, but it will be sufficient to indicate here the way in which it deals with one of them—say the move 14–18.

---

* All binary numbers are written in the convention used for the Manchester machine, i.e. with their *least* significant digit on the left.

† $W\&K$ stands for the logical product of W and K (sometimes also known as the result of collating W and K). $\sim W$ stands for the negation of W, i.e. the word obtained by writing 1's for 0's in W, and vice versa (see Chapter 15).

This type of move, which consists of adding 4 to the number of the square, corresponds to multiplying the appropriate digit in the position word by 2^4. A move of this type can be made by any black man, but only by a white king; it cannot be made from squares 28, 29, 30 or 31 nor can it be made unless the square to which the man is to be moved is empty. For a black move, the machine therefore forms the following quantity—

$$\Upsilon = \{(B\&M) \times 2^4\} \,\&\sim W\&\sim B$$

where $M = 1111, 1111, 1111, 1111, 1111, 1111, 1111, 0000$

For a white move, the corresponding quantity is—

$$\{(W\&K\&M) \times 2^4\} \,\&\sim W\&\sim B$$

In these expressions $(B\&M)$ or $(W\&K\&M)$ give all the men on the board who could make the move; multiplying this by 2^4 give the squares to which they would move. If these squares are empty (collate with $\sim W\&\sim B$) the move is possible.   [[22]]

The quantity Υ thus represents all the possible moves of this type. To consider a single one of these, the largest non-zero digit of Υ is taken and removed from Υ. The word consisting of this single digit known as θ, gives the square to which the man is moved. The quantity $\phi = \theta \times 2^{-4}$ is then formed and gives the square from which the man was moved. For a black move, the quantity—

$$B' = B \not\equiv \theta \not\equiv \phi$$ [[23]]

will then give the new position of the black men. If $K\&\phi$ is not zero, the man moved was a king so that $K' = K \not\equiv \theta \not\equiv \phi$ gives the new position of the kings. If $K\&\phi$ is zero, the man moved was not a king. The new position of the kings will therefore be unaltered unless the man has kinged during this move—in other words unless $\theta \geqslant 2^{28}$ in which case $K' = K \not\equiv \theta$.

Relatively simple modifications of this scheme are needed to deal with white moves and non-capture moves of other types. Capture moves are somewhat more complicated as multiple captures must be allowed for. Furthermore, all the possible captures must be made or the machine will render itself liable to be huffed. This leads to a considerable complication which it is not possible to describe fully here, but the basic scheme is not altered.

The machine considers all the possible moves of one type before starting the next, so that in order to describe a position fully, it is necessary to store the word Y, which indicates the moves still to be considered, as well as the position words B, W and K. It is also necessary to keep a record of the type of move being considered. This is done with the aid of a further parameter word P which also contains the value associated with the position. The whole storage required for a position is thus reduced to the 5 thirty-two-digit words B, W, K, Y, and P.

VALUATION OF POSITIONS AND STRATEGY

It should be possible to graft almost any type of strategy on to the move-finding scheme outlined above to produce a complete draughts-playing routine and then to evaluate the effectiveness of the strategy by direct experiment. I have done this with two rather simple types of strategy so far, and I hope to be able to try some rather more refined strategies in the future.

For demonstration purposes, and also to ensure that a record of the game is kept, and to take certain precautions against machine error, the move-finding sequence and the associated strategy have been combined with a general game-playing routine which accepts the opponent's moves, displays the positions, prints the move, and generally organizes the sequence of operations in the game. It is rather typical of logical programmes that this organizing routine is in fact longer than the game-playing routine proper. As its operations, though rather spectacular, are of only trivial theoretical interest, I shall not describe them here.

The first, and simplest, strategy to try is the direct one of allowing the machine to consider all the possible moves ahead on both sides for a specified number of stages. It then makes its choice, valuing the final resulting positions only in terms of the material left on the board and ignoring any positional advantage. There is an upper limit to the number of stages ahead that can be considered owing to limitations of storage space—actually six moves, three on each side, are all that can be allowed. In practice, however, time considerations provide a more severe limitation. There are on an average about ten possible legal moves at each stage of the game, so that consideration of one further stage multiplies the time for making the move by a factor of about ten. The machine considers moves at the rate of about ten a second, so that looking three moves ahead (two of its own and one of its opponents), which takes between one and two minutes, represents about the limit which can be allowed from the point of view of time.

This is not sufficient to allow the machine to play well, though it can play fairly sensibly for most of the game. One wholly unexpected difficulty appears. Consider the position on the following board.

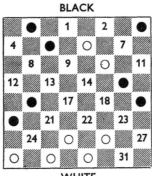

In this position, the machine is aware that its opponent is going to king next move. Now a king is more valuable than a man—the actual values used were three for a king and one for a man—so that if the opponent kings the machine effectively loses two points. The only way it can stop this is by offering a man for sacrifice, because then, by the rules of the game, the sacrifice must be taken at once. If the machine does this, it will lose only one point, and as it is not looking far enough ahead, it cannot see that it has not prevented its opponent from kinging but only postponed the evil day. At its next move it is still faced with the same difficulty, which it tries to solve in the same way, so that it will make every possible sacrifice of a single man before it accepts as inevitable the creation of an opponent's king. In fact, when faced with this position, the machine played 19—23, followed by 16—21 and 20—24.　　[24]

This, of course, is a fatal flaw in the strategy—and not one it would have been easy to discover without actually trying it out. An opponent who detected this behaviour—and it is extremely conspicuous in play—would only have to leave his man on the point of kinging indefinitely. The machine would then sacrifice all its remaining men as soon as the opportunity offered.

In order to avoid this difficulty, the second strategy was devised. In this the machine continues to investigate the moves ahead until it has found two consecutive moves without captures. This means that it will be able to recognize the futility of its sacrifice to prevent kinging. It is still necessary to impose an over-riding limit on the number of stages it can consider, and once more, considerations of time limit this. However, as no move is continued for more than two stages unless it leads to a capture, it is possible to allow the machine to consider up to four stages ahead without it becoming intolerably slow. This would mean that it would consider the sacrifice of two men to be of equal value to the creation of an

	Machine	Strachey
1.	11—15	23—18
2.	7—11	21—17
3.	8—12	20—16[1]
4.	12—21(16)	25—16(21)
5.	9—14![2]	18—9(14)
6.	6—20(16,9)[3]	27—23
7.	2—7[4]	23—18
8.	5—8	18—14
9.	8—13[5]	17—8(13)
10.	4—13(8)	14—9
11.	1—5[6]	9—6
12.	15—19	6—1 K
13.	5—9	1—6?[7]
14.	0—5![8]	6—15(10)
15.	11—25(22,15)	30—21(25)
16.	13—17	21—14(17)
17.	9—18(14)	24—21
18.	18—23	26—22
19.	23—27	22—17
20.	5—8[9]	17—14
21.	8—13	14—9
22.	19—23	9—6
23.	23—26[10]	31—22(26)
24.	27—31 K	6—2 K
25.	7—10	2—7
26.	10—15	21—16?[11]
27.	3—10(7)	16—9(13)
28.	10—14	9—6
29.	15—19	6—2 K
30.	31—27[12]	2—6
31.	27—31[12]	6—10
32.	31—26[13]	10—17(14)
33.	19—23	29—25
34.	26—31[14]	

Notes—

1. An experiment on my part—the only deliberate offer I made. I thought, wrongly, that it was quite safe.
2. Not foreseen by me.
3. Better than 5—21(9,17).
4. A random move (zero value). Shows the lack of a constructive plan.
5. Another random move of zero value, actually rather good.
6. Bad. Ultimately allows me to make a King. 10—14 would have been better.
7. A bad slip on my part.
8. Taking full advantage of my slip.
9. Bad. Unblocks the way to a King.
10. Sacrifice in order to get a King (not to stop me kinging). A good move, but not possible before 19—23 had been made by chance.
11. Another bad slip on my part.
12. Purposeless. The strategy is failing badly in the end game.
13. Too late.
14. Futile. The game was stopped at this point as the outcome was obvious.

opponent's king, and as there is a random choice between moves of equal value, it might still make this useless sacrifice. This has been prevented by reducing the value of a king from 3 to $2\frac{7}{8}$.

With this modified strategy, the machine can play quite a tolerable game until it reaches the end game. It has always seemed probable that a wholly different strategy will be necessary for end games. The game given on page 303, which is the first ever played using the strategy, brings this point out very clearly. ⟦25⟧

NIM

A considerably easier game which the machine can be programmed to play is the one known as nim. Probably a variation of this was known to the Chinese—certainly in its present form many people have met it. We have chosen to deal with this comparatively trivial game in detail because of its topical interest. Thousands of people will have seen *Nimrod*, the computer built by Ferranti Ltd. for the Science Exhibition of the Festival of Britain. This special-purpose machine was designed to show the main features of large electronic digital computers, and the game of nim was chosen as an interesting but simple demonstration problem. The game itself is as follows—

Initially we have any number of heaps, each containing any number of tokens (usually matches). In the simplest form, two contestants play alternately, and may pick up as many matches as they wish at one time from *one* pile, but they must take at least one match. The aim is to avoid taking the last match of all—or there is another variation where the aim is to take the last match or group of matches.

The so-called *multiple game* differs from this only in that the number of heaps altered in any move may take any value from one up to a pre-assigned maximum k. Of course, to prevent complete triviality, k must be less than N, the total number of heaps.

The detailed theory of nim was worked out long ago and, apart from the initial distribution of the matches, no element of chance need enter into the game. This theory is very simple, but it becomes clearer for the non-mathematician if we use the concept of a binary number, introduced elsewhere (see page 33).

We can now proceed to give a working rule for the game of nim. We would like to find a *winning position* having the following characteristics—

(*a*) It is impossible, when faced by a winning position, to make a move which will leave a winning position.

(b) Faced with any other than a winning position, it is possible to make a move resulting in a winning position.

(c) If at any stage of the game a player A can convert a position into a winning position, it is possible for A to win, and impossible for his opponent B to do so unless A makes a mistake. A wins by leaving a winning position at every succeeding move on his part.

Such winning positions can be achieved and are recognized as follows: For any given configuration, express the number of matches in each heap as a binary number. Suppose, for example, that we have four heaps, A, B, C and D, containing respectively 7, 4, 3 and 2 matches. These are represented—

	4	2	1	
A	1	1	1	(7)
B	1	0	0	(4)
C	0	1	1	(3)
D	0	1	0	(2)

We write these down as above, one under the other, and add up each column, e.g., in the above example, we get

	4	2	1
Sum:	2	3	2

Now the "secret" of a winning position is that every column should be divisible by $k + 1$; k being the maximum number of heaps which can be altered in any one move. Thus the example quoted above cannot represent a winning position whatever our initial choice of k. However, suppose we have $k = 1$; then consider the position—

	4	2	1	
A	1	0	1	(5)
B	1	1	1	(7)
C	0	1	1	(3)
D	0	0	1	(1)
Sum:	2	2	4	

This is a winning position, but would not be so if we had previously fixed $k = 3$, for example.

To convert an "unsafe" into a winning position, we deal with a column at a time. Consider our previous example with $k = 1$.

	4	2	1	
A	1	1	1	(7)
B	1	0	0	(4)
C	0	1	1	(3)
D	0	1	0	(2)
Sum:	2	3	2	

We start with the "most-significant," or left-hand column. This sum is divisible by $k + 1$, so we proceed to consideration of the next column. The sum here is 3, which is not divisible by $k + 1$, so we choose any heap, say D, having a one in this column. We remove this 1 (which is equivalent to subtracting 2 from D), and put 1 in every less-significant (or right-hand) column of this heap (which in this case is equivalent to adding 1, though if we had chosen to modify A instead, it would have meant no change in the last column). That is, we make the minimum move which removes the 1 in the "unsafe" column. Thus we remove 1 from D, and so alter its binary representation to 001.

Now our representation is—

	4	2	1	
A	1	1	1	(7)
B	1	0	0	(4)
C	0	1	1	(3)
D	0	0	1	(1)
Sum:	2	2	3	

and we see that we have made the sum of column 2 divisible by $k + 1$ at the expense of column 1. However, we shall now proceed to adjust column 1. To avoid altering more than k heaps in one move, we must alter one or more of the heaps already affected if, by so doing, we can achieve the desired result, rather than select a fresh heap.

[[26]] Now, in this case, we wish to remove 1 from column 1 of some heap. Since heap D has already been altered, we choose this—it has a 1 in this column.

So, at the end of our move, we have removed two matches from heap D, and leave the winning position—

	4	2	1	
A	1	1	1	(7)
B	1	0	0	(4)
C	0	1	1	(3)
D	0	.0	0	(0)
Sum:	2	2	2	

In adapting this game for the universal computer, we allow a maximum of eight heaps, with not more than thirty-one matches in a heap. In Nimrod the more stringent restrictions to four heaps, each with a maximum content of seven matches, were applied to simplify the problems of demonstration.

Possible positions with which the machine may be faced are as follows—

(a) At least $k + 1$ heaps contain more than one match.

(b) The number of heaps containing more than one match lies between 1 and k (inclusive).

(c) No heap contains more than one match. Not all heaps are empty.

(d) All heaps are empty.

In case (a), we follow the so-called *normal routine*, which aims at leaving column sums all divisible by $(k + 1)$.

In case (b), we want to leave $r (k + 1) + 1$ heaps containing one match, and no heaps with more than one, where r may have any non-negative integral value (i.e. $r = 0, 1, 2, \ldots$).

In case (c) the same applies. If only one heap is left, containing one match, we have no choice of move, but this need not be treated separately.

In case (d), the game is over. Special investigation has to be used to detect this case. In all other cases, if the normal routine cannot succeed in its purpose, i.e. if the machine is faced with a winning position—a random move can, and must, be made. But, in this situation, this obviously cannot be done.

Thus the routine breaks up naturally into the following parts—

   (i) Entry
  (ii) Determination of case
 (iii) Normal Routine
  (iv) Cases (*b*) and (*c*)
   (v) Treatment of zero case (*d*)
  (vi) Random move
 (vii) Emergence.

There is no need to give further details of the programme, but an example is given of how the machine would tackle a specific game.

Suppose initially that we have four heaps, containing respectively 7, 4, 5 and 2 matches; that $k = 2$; and that the machine moves first.

   (i) Entry—

	4	2	1	
A	1	1	1	(7)
B	1	0	0	(4)
C	1	0	1	(5)
D	0	1	0	(2)

  (ii) Determination of case—

There are 4 non-zero, non-unit heaps, so we are dealing with case (*a*).

 (iii) Normal routine—

	4	2	1	
A	1	1	1	(7)
B	1	0	0	(4)
C	1	0	1	(5)
D	0	1	0	(2)
Sum:	3	2	2	

The sum of column 4 is divisible by $k + 1$ so we need not modify it.

The sum of column 2 is 2, and is not divisible by $k + 1$, so we need to modify any heap having a 1 in this column—say heap A.

According to the rules, we then get—

	4	2	1	
A	1	0	1	(5)
B	1	0	0	(4)
C	1	0	1	(5)
D	0	1	0	(2)
Sum:	3	1	2	

And we note that heap A has been modified, and should be again modified whenever possible. Sum of column 2 is still not divisible by $k + 1$, so this time we modify heap D to obtain—

	4	2	1	
A	1	0	1	(5)
B	1	0	0	(4)
C	1	0	1	(5)
D	0	0	1	(1)
Sum:	3	0	3	

Column 2 is now divisible by $k + 1$ and, proceeding to the next column, we see this condition is also satisfied here, so the move has been completed and a winning-position left, the means to this end being the removal of two matches from A, and one from D, leaving 5, 4, 5 and 1. (If column 1 had needed adjustment, we should have had to modify one or both of heaps A and D, since these had already been affected.)

Suppose the opponent now makes a move leaving 0, 4, 2 and 1 as the contents of the respective heaps. It is now for the machine to move again.

(i) Entry—

	4	2	1	
A	0	0	1	(0)
B	1	0	0	(4)
C	0	1	0	(2)
D	0	0	1	(1)

(ii) Determination of case.

There are 3 non-zero, non-unit heaps, so we are dealing with case (b). Thus we want to leave 1, or 4, or 7 . . . unit-heaps. Clearly we can only leave 1 unit heap in this case.

(iv) Cases (b) and (c).

We remove all matches from heaps B and D, which affects only k heaps, and leaves just one unit heap as required.

The opponent is now forced to remove the last match, and the machine wins the game.

SOLVABLE AND UNSOLVABLE PROBLEMS

A. M. TURING, F. R. S.

IF one is given a puzzle to solve one will usually, if it proves to be difficult, ask the owner whether it can be done. Such a question should have a quite definite answer, yes or no, at any rate provided the rules describing what you are allowed to do are perfectly clear. Of course the owner of the puzzle may not know the answer. One might equally ask, 'How can one tell whether a puzzle is solvable?', but this cannot be answered so straightforwardly. The fact of the matter is that there is *no* systematic method of testing puzzles to see whether they are solvable or not. If by this one meant merely that nobody had ever yet found a test which could be applied to any puzzle, there would be nothing at all remarkable in the statement. It would have been a great achievement to have invented such a test, so we can hardly be surprised that it has never been done. But it is not merely that the test has never been found. It has been proved that no such test ever can be found.

Let us get away from generalities a little and consider a particular puzzle. One which has been on sale during the last few years and has probably been seen by most of the readers of this article illustrates a number of the points involved quite well. The puzzle consists of a large square within which are some smaller movable squares numbered 1 to 15, and one empty space, into which any of the neighbouring squares can be slid leaving a new empty space behind it. One may be asked to transform a given arrangement of the squares into another by a succession of such movements of a square into an empty space. For this puzzle there is a fairly simple and quite practicable rule by which one can tell whether the transformation required is possible or not. One first imagines the transformation carried out according to a different set of rules. As well as sliding the squares into the empty space one is allowed to make moves each consisting of two interchanges, each of one pair of squares. One would, for instance, be allowed as one move to interchange the squares numbered 4 and 7, and also the squares numbered 3 and 5. One is permitted to use the same number in both pairs. Thus one may replace 1 by 2, 2 by 3, and 3 by 1 as a move because this is the same as interchanging first (1, 2) and then (1, 3). The original puzzle is solvable by sliding if it is solvable according to the new rules. It is not solvable by sliding if the required position can be reached by the new rules, together with a 'cheat' consisting of *one single* interchange of a pair of squares.* Suppose, for instance, that one is asked to get back to the standard position –

1	2	3	4
5	6	7	8
9	10	11	12
13	14	15	////

from the position

10	1	4	5
9	2	6	8
11	3	////	15
13	14	7	12

One may, according to the modified rules, first get the empty square into the correct position by moving the squares 15 and 12, and then get the squares 1, 2, 3, ... successively into their correct positions by the interchanges (1, 10), (2, 10), (3, 4), (4, 5), (5, 9), (6, 10), (7, 10), (9, 11), (10, 11), (11, 15). The squares 8, 12, 13, 14, 15 are found to be already in their correct positions when their turns are reached. Since the number of interchanges required is

---

*It would take us too far from our main purpose to give the proof of this rule: the reader should have little difficulty in proving it by making use of the fact that an odd number of interchanges can never bring a set of objects back to the position it started from.

even, this transformation is possible by sliding.† If one were required after this to interchange say square 14 and 15 it could not be done.   [[1]]

This explanation of the theory of the puzzle can be regarded as entirely satisfactory. It gives one a simple rule for determining for any two positions whether one can get from one to the other or not. That the rule is so satisfactory depends very largely on the fact that it does not take very long to apply. No mathematical method can be useful for any problem if it involves much calculation. It is nevertheless sometimes interesting to consider whether something is possible at all or not, without worrying whether, in case it *is* possible, the amount of labour or calcu-   [[2]] lation is economically prohibitive. These investigations that are not concerned with the amount of work involved are in some ways easier to carry out, and they certainly have a greater aesthetic appeal. The results are not altogether without value, for if one has proved that there is no method of doing something it follows *a fortiori* that there is no practicable method. On the other hand, if one method has been proved to exist by which the decision can be made, it gives some encouragement to anyone who wishes to find a workable method.

From this point of view, in which one is only interested in the question, 'Is there a systematic way of deciding whether puzzles of this kind are solvable?', the rules which have been described for the sliding-squares puzzle are much more special and detailed than is really necessary. It would be quite enough to say: 'Certainly one can find out whether one position can be reached from another by a systematic procedure. There are only a finite number of positions in which the numbered squares can be arranged (viz. 20922789888000) and only a finite number (2, 3, or 4) of moves in each position. By making a list of all the

†It can in fact be done by sliding successively the squares numbered 7, 14, 13, 11, 9, 10, 1, 2, 3, 7, 15, 8, 5, 4, 6, 3, 10, 1, 2, 6, 3, 10, 6, 2, 1, 6, 7, 15, 8, 5, 10, 8, 5, 10, 8, 7, 6, 9, 15, 5, 10, 8, 7, 6, 5, 15, 9, 5, 6, 7, 8, 12, 14, 13, 15, 10, 13, 15, 11, 9, 10, 11, 15, 13, 12, 14, 13, 15, 9, 10, 11, 12, 14, 13, 15, 14, 13, 15, 14, 13, 12, 11, 10, 9, 13, 14, 15, 12, 11, 10, 9, 13, 14, 15.

positions and working through all the moves, one can divide the
positions into classes, such that sliding the squares allows one to
get to any position which is in the same class as the one started
from. By looking up which classes the two positions belong to
one can tell whether one can get from one to the other or not.'
This is all, of course, perfectly true, but one would hardly find
such remarks helpful if they were made in reply to a request for
an explanation of how the puzzle should be done. In fact they
are so obvious that under such circumstances one might find
them somehow rather insulting. But the fact of the matter is,
that if one is interested in the question as put, 'Can one tell by
a systematic method in which cases the puzzle is solvable?', this
answer is entirely appropriate, because one wants to know if
there is a systematic method, rather than to know of a good one.

The same kind of argument will apply for any puzzle where
one is allowed to move certain 'pieces' around in a specified
manner, provided that the total number of essentially different
positions which the pieces can take up is finite. A slight variation
on the argument is necessary in general to allow for the fact that
in many puzzles some moves are allowed which one is not permitted to reverse. But one can still make a list of the positions,
and list against these first the positions which can be reached
from them in one move. One then adds the positions which are
reached by two moves and so on until an increase in the number
of moves does not give rise to any further entries. For instance,
we can say at once that there is a method of deciding whether a
patience can be got out with a given order of the cards in the
pack: it is to be understood that there is only a finite number of
places in which a card is ever to be placed on the table. It may
be argued that one is permitted to put the cards down in a
[3]   manner which is not perfectly regular, but one can still say that
there is only a finite number of 'essentially different' positions.
A more interesting example is provided by those puzzles made
(apparently at least) of two or more pieces of very thick twisted
wire which one is required to separate. It is understood that one
is not allowed to bend the wires at all, and when one makes the
right movement there is always plenty of room to get the pieces

apart without them ever touching, if one wishes to do so. One may describe the positions of the pieces by saying where some three definite points of each piece are. Because of the spare space it is not necessary to give these positions quite exactly. It would be enough to give them to, say, a tenth of a millimetre. One does not need to take any notice of movements of the puzzle as a whole: in fact one could suppose one of the pieces quite fixed. The second piece can be supposed to be not very far away, for, if it is, the puzzle is already solved. These considerations enable us to reduce the number of 'essentially different' positions to a finite number, probably a few hundred millions, and the usual argument will then apply. There are some further complications, which we will not consider in detail, if we do not know how much clearance to allow for. It is necessary to repeat the process again and again allowing successively smaller and smaller clearances. Eventually one will find that either it can be solved, allowing a small clearance margin, or else it cannot be solved even allowing a small margin of 'cheating' (i.e. of 'forcing', or having the pieces slightly overlapping in space). It will, of course, be understood that this process of trying out the possible positions is not to be done with the physical puzzle itself, but on paper, with mathematical descriptions of the positions, and mathematical criteria for deciding whether in a given position the pieces overlap, etc.

These puzzles where one is asked to separate rigid bodies are in a way like the 'puzzle' of trying to undo a tangle, or more generally of trying to turn one knot into another without cutting the string. The difference is that one is allowed to bend the string, but not the wire forming the rigid bodies. In either case, if one wants to treat the problem seriously and systematically one has to replace the physical puzzle by a mathematical equivalent. The knot puzzle lends itself quite conveniently to this. A knot is just a closed curve in three dimensions nowhere crossing itself; but, for the purpose we are interested in, any knot can be given accurately enough as a series of segments in the directions of the three coordinate axes. Thus, for instance, the trefoil knot (Figure 1*a*) may be regarded as consisting of a number of segments joining the points given, in the usual (x, y, z) system of coordinates, as $(1, 1, 1)$, $(4, 1, 1,)$, $(4, 2, 1)$, $(4, 2, -1)$, $(2, 2, -1)$, $(2, 2, 2)$, $(2, 0, 2)$, $(3, 0, 2)$, $(3, 0, 0)$, $(3, 3, 0)$, $(1, 3, 0)$, $(1, 3, 1)$ and returning again with a twelfth segment to the starting point $(1, 1, 1)$. This representation of the knot is shown in perspective in Figure 1*b*. There is no special virtue in the representation which has been chosen. If it is desired to follow the original curve more closely a greater number of segments must be used. Now let *a* and *d* represent unit steps in the positive and negative X-directions respectively, *b* and *e* in the Y-directions, and *c* and *f* in the Z-directions: then this knot may be described as *aaabffddcccceeaffbbbbddcee*. One can then, if one wishes, deal entirely with such sequences of letters. In order that such a sequence should represent a knot it is necessary and sufficient that the numbers of *a*'s and *d*'s should be equal, and likewise the number of *b*'s equal to the number of *e*'s and the number of *c*'s equal to the number of *f*'s, and it must not be possible to obtain another sequence of letters with these properties by omitting a number of consecutive letters at the beginning or the end or both. One can turn a knot into an equivalent one by operations of the following kinds—

  (i) One may move a letter from one end of the row to the other.
 (ii) One may interchange two consecutive letters provided this still gives a knot.
(iii) One may introduce a letter *a* in one place in the row, and *d* somewhere else, or *b* and *e,* or *c* and *f,* or take such pairs out, provided it still gives a knot.
 (iv) One may replace *a* everywhere by *aa* and *d* by *dd* or replace each *b* and *e* by *bb* and *ee* or each *c* and *f* by *cc* and *ff*. One may also reverse any such operation.

—and these are all the moves that are necessary.

It is also possible to give a similar symbolic equivalent for the problem of separating rigid bodies, but it is less straightforward than in the case of knots.

These knots provide an example of a puzzle where one cannot tell in advance how many arrangements of pieces may be involved (in this case the pieces are the letters *a, b, c, d. e, f*), so that

(a)

(b)

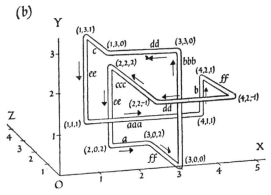

Fig. 1. (a) The trefoil knot (b) a possible representation of this knot as a number of segments joining points.

the usual method of determining whether the puzzle is solvable cannot be applied. Because of rules (iii) and (iv) the lengths of the sequences describing the knots may become indefinitely great. No systematic method is yet known by which one can tell whether two knots are the same.

Another type of puzzle which we shall find very important is the 'substitution puzzle'. In such a puzzle one is supposed to be supplied with a finite number of different kinds of counters, perhaps just black (B) and white (W). Each kind is in unlimited supply. Initially a number of counters are arranged in a row and one is asked to transform it into another pattern by substitutions. A finite list of the substitutions allowed is given. Thus, for instance, one might be allowed the substitutions

$$\text{(i)} \quad WBW \rightarrow B$$
$$\text{(ii)} \quad BW \rightarrow WBBW$$

and be asked to transform WBW into $WBBBW$, which could be done as follows

$$WBW \to WWBBW \to WWBWBBW \to WBBBW$$
$$\underline{\quad}\text{(ii)} \quad \underline{\quad\quad}\text{(ii)} \quad \underline{\quad\quad}\text{(i)}$$

Here the substitutions used are indicated by the numbers below the arrows, and their effects by underlinings. On the other hand if one were asked to transform WBB into BW it could not be done, for there are no admissible steps which reduce the number of B's.

It will be seen that with this puzzle, and with the majority of substitution puzzles, one cannot set any bound to the number of positions that the original position might give rise to.

It will have been realized by now that a puzzle can be something rather more important than just a toy. For instance the task of proving a given mathematical theorem within an axiomatic system is a very good example of a puzzle.

It would be helpful if one had some kind of 'normal form' or 'standard form' for describing puzzles. There is, in fact, quite a reasonably simple one which I shall attempt to describe. It will be necessary for reasons of space to take a good deal for granted, but this need not obscure the main ideas. First of all we may suppose that the puzzle is somehow reduced to a mathematical form in the sort of way that was used in the case of the knots. The position of the puzzle may be described, as was done in that case, by sequences of symbols in a row. There is usually very little difficulty in reducing other arrangements of symbols (e.g. the squares in the sliding squares puzzle) to this form. The question which remains to be answered is, 'What sort of rules should one be allowed to have for rearranging the symbols or counters?' In order to answer this one needs to think about what kinds of processes ever do occur in such rules, and, in order to reduce their number, to break them up into simpler processes. Typical of such processes are counting, copying, comparing, substituting. When one is doing such processes, it is necessary, especially if there are many symbols involved, and if one wishes to avoid carrying too much information in one's head, either to make a number of jottings elsewhere or to use a number of

[4]

[194]

marker objects as well as the pieces of the puzzle itself. For instance, if one were making a copy of a row of counters concerned in the puzzle it would be as well to have a marker which divided the pieces which have been copied from those which have not and another showing the end of the portion to be copied. Now there is no reason why the rules of the puzzle itself should not be expressed in such a way as to take account of these markers. If one does express the rules in this way they can be made to be just substitutions. This means to say that the *normal form for puzzles is the substitution type of puzzle.* More definitely we can say:

Given any puzzle we can find a corresponding substitution puzzle *which is equivalent to it in the sense that given a solution of the one we can easily use it to find a solution of the other. If the original puzzle is concerned with rows of pieces of a finite number of different kinds, then the substitutions may be applied as an alternative set of rules to the pieces of the original puzzle. A transformation can be carried out by the rules of the original puzzle if and only if it can be carried out by the substitutions and leads to a final position from which all marker symbols have disappeared.*

This statement is still somewhat lacking in definiteness, and will remain so. I do not propose, for instance, to enter here into the question as to what I mean by the word 'easily'. The statement is moreover one which one does not attempt to prove. Propaganda is more appropriate to it than proof, for its status is something between a theorem and a definition. In so far as we know *a priori* what is a puzzle and what is not, the statement is a theorem. In so far as we do not know what puzzles are, the statement is a definition which tells us something about what they are. One can of course define a puzzle by some phrase beginning, for instance, 'A set of definite rules . . .', but this just throws us back on the definition of 'definite rules'. Equally one can reduce it to the definition of 'computable function' or 'systematic procedure'. A definition of any one of these would define all the rest. Since 1935 a number of definitions have been given, explaining in detail the meaning of one or other of these terms, and these have all been proved equivalent to one another and also equivalent to the above statement. In effect there is no opposition to the view that every puzzle is equivalent to a substitution puzzle.

After these preliminaries let us think again about puzzles as a whole. First let us recapitulate. There are a number of questions to which a puzzle may give rise. When given a particular task one may ask quite simply

(a) *Can this be done?*

Such a straightforward question admits only the straightforward answers, 'Yes' or 'No', or perhaps 'I don't know'. In the case that the answer is 'Yes' the answerer need only have done the puzzle himself beforehand to be sure. If the answer is to be 'No', some rather more subtle kind of argument, more or less mathematical, is necessary. For instance, in the case of the sliding squares one can state that the impossible cases *are* impossible because of the mathematical fact that an odd number of simple interchanges of a number of objects can never bring one back to where one started. One may also be asked

(b) *What is the best way of doing this?*

Such a question does not admit of a straightforward answer. It depends partly on individual differences in people's ideas as to what they find easy. If it is put in the form, 'What is the solution which involves the smallest number of steps?', we again have a straightforward question, but now it is one which is somehow of remarkably little interest. In any particular case where the answer to (a) is 'Yes' one can find the smallest possible number of steps by a tedious and usually impracticable process of enumeration, but the result hardly justifies the labour.

When one has been asked a number of times whether a number of different puzzles of similar nature can be solved one is naturally led to ask oneself

(c) *Is there a systematic procedure by which I can answer these questions, for puzzles of this type?*

If one were feeling rather more ambitious one might even ask

(d) *Is there a systematic procedure by which one can tell whether a puzzle is solvable?*

I hope to show that the answer to this last question is 'No'.

[5]

There are in fact certain types of puzzle for which the answer to (c) is 'No'.

Before we can consider this question properly we shall need to be quite clear what we mean by a 'systematic procedure' for deciding a question. But this need not now give us any particular difficulty. A 'systematic procedure' was one of the phrases which we mentioned as being equivalent to the idea of a puzzle, because either could be reduced to the other. If we are now clear as to what a puzzle is, then we should be equally clear about 'systematic procedures'. In fact a systematic procedure is just a puzzle *in which there is never more than one possible move in any of the positions which arise and in which some significance is attached to the final result.*

Now that we have explained the meaning both of the term 'puzzle' and of 'systematic procedure', we are in a position to prove the assertion made in the first paragraph of this article, that there cannot be any systematic procedure for determining whether a puzzle be solvable or not. The proof does not really require the detailed definition of either of the terms, but only the relation between them which we have just explained. Any systematic procedure for deciding whether a puzzle were solvable could certainly be put in the form of a puzzle, with unambiguous moves (i.e. only one move from any one position), and having for its starting position a combination of the rules, the starting position and the final position of the puzzle under investigation.

The puzzle under investigation is also to be described by its rules and starting position. Each of these is to be just a row of symbols. As we are only considering substitution puzzles, the rules need only be a list of all the substitution pairs appropriately punctuated. One possible form of punctuation would be to separate the first member of a pair from the second by an arrow, and to separate the different substitution pairs with colons. In this case the rules

B may be replaced by *BC*

WBW may be deleted

would be represented by ' $: B \to BC : WBW \to :$ ' . For the purposes of the argument which follows, however, these arrows and colons are an embarrassment. We shall need the rules to be expressed without the use of any symbols which are barred from appearing in the starting positions. This can be achieved by the following simple, though slightly artificial trick. We first double all the symbols other than the punctuation symbols, thus ': $BB \rightarrow BBCC$: $WWBBWW \rightarrow$:' . We then replace each arrow by a single symbol, which must be different from those on either side of it, and each colon by three similar symbols, also chosen to avoid clashes. This can always be done if we have at least three symbols available, and the rules above could then be represented as, for instance, '$CCCBBWBBCCBBBWWBBWW$ $BWWW$'. Of course according to these conventions a great variety of different rows of symbols will describe essentially the same puzzle. Quite apart from the arbitrary choice of the punctuating symbols the substitution pairs can be given in any order, and the same pair can be repeated again and again.

Now let $P(R,S)$ stand for 'the puzzle whose rules are described by the row of symbols R and whose starting position is described by S'. Owing to the special form in which we have chosen to describe the rules of puzzles, there is no reason why we should not consider $P(R,R)$ for which the 'rules' also serve as starting position : in fact the success of the argument which follows depends on our doing so. The argument will also be mainly concerned with puzzles in which there is at most one possible move in any position; these may be called 'puzzles with un-ambiguous moves'. Such a puzzle may be said to have 'come out' if one reaches either the position B or the position W, and the rules do not permit any further moves. Clearly if a puzzle has unambiguous moves it cannot both come out with the end result B and with the end result W.

We now consider the problem of classifying rules R of puzzles into two classes, I and II, as follows:

Class I is to consist of sets R of rules, which represent puzzles with unambiguous moves, and such that $P(R,R)$ comes out with the end result W.

Class II is to include all other cases, i.e. either $P(R,R)$ does

[[6]]

not come out, or comes out with the end result *B*, or else *R* does not represent a puzzle with unambiguous moves. We may also, if we wish, include in this class sequences of symbols such as *BBBBB* which do not represent a set of rules at all.

Now suppose that, contrary to the theorem that we wish to prove, we have a systematic procedure for deciding whether puzzles come out or not. Then with the aid of this procedure we shall be able to distinguish rules of class I from those of class II. There is no difficulty in deciding whether *R* really represents a set of rules, and whether they are unambiguous. If there is any difficulty it lies in finding the end result in the cases where the puzzle is known to come out: but this can be decided by actually working the puzzle through. By a principle which has already been explained, this systematic procedure for distinguishing the two classes can itself be put into the form of a substitution puzzle (with rules *K*, say). When applying these rules *K*, the rules *R* of the puzzle under investigation form the starting position, and the end result of the puzzle gives the result of the test. Since the procedure always gives an answer, the puzzle *P(K,R)* always comes out. The puzzle *K* might be made to announce its results in a variety of ways, and we may be permitted to suppose that the end result is *B* for rules *R* of class I, and *W* for rules of class II. The opposite choice would be equally possible, and would hold for a slightly different set of rules *K'*, which however we do not choose to favour with our attention. The puzzle with rules *K* may without difficulty be made to have unambiguous moves. Its essential properties are therefore:

[[7]]

> *K* has unambiguous moves.
> *P(K,R)* always comes out whatever *R*.
> If *R* is in class I, then *P(K,R)* has end result *B*.
> If *R* is in class II, then *P(K,R)* has end result *W*.

These properties are however inconsistent with the definitions of the two classes. If we ask ourselves which class *K* belongs to, we find that neither will do. The puzzle *P(K,K)* is bound to come out, but the properties of *K* tell us that we must get end result *B* if *K* is in class I and *W* if it is in class II, whereas the definitions of the classes tell us that the end results must be the other way round. The assumption that there was a systematic procedure for telling whether puzzles come out has thus been reduced to an absurdity.

Thus in connexion with question (c) above we can say that there are some types of puzzle for which no systematic method of deciding the question exists. This is often expressed in the form, 'There is no *decision procedure* for this type of puzzle', or again, 'The decision problem for this type of puzzle is unsolvable', and so one comes to speak (as in the title of this article) about 'unsolvable problems' meaning in effect puzzles for which there is no decision procedure. This is the technical meaning which the words are now given by mathematical logicians. It would seem more natural to use the phrase 'unsolvable problem' to mean just an unsolvable puzzle, as for example 'to transform 1, 2, 3 into 2, 1, 3 by cyclic permutation of the symbols', but this is not the meaning it now has. However, to minimize confusion I shall here always speak of 'unsolvable decision problems', rather than just 'unsolvable problems', and also speak of puzzles rather than problems where it is puzzles and not decision problems that are concerned.

It should be noticed that a decision problem only arises when one has an infinity of questions to ask. If you ask, 'Is this apple good to eat?', or 'Is this number prime?', or 'Is this puzzle solvable?' the question can be settled with a single 'Yes' or 'No'. A finite number of answers will deal with a question about a finite number of objects, such as the apples in a basket. When the number is infinite, or in some way not yet completed concerning say all the apples one may ever be offered, or all whole numbers or puzzles, a list of answers will not suffice. Some kind of rule or systematic procedure must be given. Even if the number concerned is finite one may still prefer to have a rule rather than a list: it may be easier to remember. But there certainly cannot be an unsolvable decision problem in such cases, because of the possibility of using finite list.

Regarding decision problems as being concerned with classes of puzzles, we see that if we have a decision method for one class it will apply also for any subclass. Likewise, if we have proved that there is no decision procedure for the subclass, it follows that there is none for the whole class. The most interesting and valuable results about unsolvable decision problems concern the smaller classes of puzzle.

Another point which is worth noticing is quite well illustrated by the puzzle which we considered first of all in which the pieces were sliding squares. If one wants to know whether the puzzle is solvable with a given starting position, one can try moving the pieces about in the hope of reaching the required end-position. If one succeeds, then one will have solved the puzzle and consequently will be able to answer the question, 'Is it solvable?' In the case that the puzzle is solvable one will eventually come on the right set of moves. If one has also a procedure by which, if the puzzle is unsolvable, one would eventually establish the fact that it was so, then one would have a solution of the decision problem for the puzzle. For it is only necessary to apply both processes, a bit of one alternating with a bit of the other, in order eventually to reach a conclusion by one or the other. Actually, in the case of the sliding squares problem, we have got such a procedure, for we know that if, by sliding, one ever reaches the required final position, with squares 14 and 15 interchanged, then the puzzle is impossible.

It is clear then that the difficulty in finding decision procedures for types of puzzle lies in establishing that the puzzle is unsolvable in those cases where it *is* unsolvable. This, as was mentioned on page 16, requires some sort of mathematical argument. This suggests that we might try expressing the statement that the puzzle comes out in a mathematical form and then try and prove it by some systematic process. There is no particular difficulty in the first part of this project, the mathematical expression of the statement about the puzzle. But the second half of the project is bound to fail, because by a famous theorem of Gödel no ⟦8⟧ systematic method of proving mathematical theorems is sufficiently complete to settle every mathematical question, yes or no. In any case we are now in a position to give an independent proof of this. If there were such a systematic method of proving mathematical theorems we could apply it to our puzzles and for each one eventually either prove that it was solvable or unsolvable; this would provide a systematic method of determining whether the puzzle was solvable or not, contrary to what we have already proved.

This result about the decision problem for puzzles, or, more accurately speaking, a number of others very similar to it, was proved in 1936–7. Since then a considerable number of further decision problems have been shown to be unsolvable. They are all proved to be unsolvable by showing that if they were solvable one could use the solution to provide a solution of the original one. They could all without difficulty be reduced to the same unsolvable problem. A number of these results are mentioned very shortly below. No attempt is made to explain the technical terms used, as most readers will be familiar with some of them, and the space required for the explanation would be quite out of proportion to its usefulness in this context.

(1) It is not possible to solve the decision problem even for substitution processes applied to rows of black and white counters only.

(2) There are certain particular puzzles for which there is no decision procedure, the rules being fixed and the only variable element being the starting position.

(3) There is no procedure for deciding whether a given set of axioms leads to a contradiction or not.

⟦9⟧ (4) The 'word problem in semi-groups with cancellation' is not solvable.

(5) It has recently been announced from Russia that the 'word problem in groups' is not solvable. This is a decision problem not unlike the 'word problem in semi-groups', but very much more important, having applications in topology: attempts were being made to solve this decision problem before any such problems had been proved unsolvable. No adequately complete proof is yet available, but if it is correct this is a considerable step forward.

(6) There is a set of 102 matrices of order 4, with integral coefficients such that there is no decision method for determining

whether another given matrix is or is not expressible as a product of matrices from the given set.

These are, of course, only a selection from the results. Although quite a number of decision problems are now known to be unsolvable, we are still very far from being in a position to say of a given decision problem, whether it is solvable or not. Indeed, we shall never be quite in that position, for the question whether a given decision problem is solvable is itself one of the undecidable decision problems. The results which have been found are on the whole ones which have fallen into our laps rather than ones which have positively been searched for. Considerable efforts have however been made over the word problem in groups (see (5) above). Another problem which mathematicians are very anxious to settle is known as 'the decision problem of the equivalence of manifolds'. This is something like one of the problems we have already mentioned, that concerning the twisted wire puzzles. But whereas with the twisted wire puzzles the pieces are quite rigid, the 'equivalence of manifolds' problem concerns pieces which one is allowed to bend, stretch, twist, or compress as much as one likes, without ever actually breaking them or making new junctions or filling in holes. Given a number of interlacing pieces of plasticine one may be asked to transform them in this way into another given form. The decision problem for this class of problem is the 'decision problem for the equivalence of manifolds'. It is probably unsolvable, but has never been proved to be so. A similar decision problem which might well be unsolvable is the one concerning knots which has already been mentioned.

The results which have been described in this article are mainly of a negative character, setting certain bounds to what we can hope to achieve purely by reasoning. These, and some other results of mathematical logic may be regarded as going some way towards a demonstration, within mathematics itself, of the inadequacy of 'reason' unsupported by common sense.

FURTHER READING

Kleene, S. C. *Introduction to Metamathematics*, Amsterdam, 1952.

NOTES

1945 *Proposal for Development in the Mathematics Division of an Automatic Computing Engine (ACE)*

[[1]]   A reference to early analogue computers which were used to solve differential equations using the non-discrete values of voltage.

[[2]]   Presumably Turing is referring to the intervention of human operators during the process of analogue computing where, for example, potentiometers would be adjusted in order for families of differential equations to be solved.

[[3]]   This statement might seem rather surprising to us today where hardware reliability is a minor problem, compared with, say, software reliability. However, one must bear in mind that the hardware technology used by Turing was still primitive and untried.

[[4]]   A description of a piece of system software known as a loader whose purpose is to clear a computer's memory and load a particular program ready for execution.

[[5]]   The mercury tanks operated as delay lines.

[[6]]   This limitation is, of course, a result of the primitive hardware technology that was available, the order codes of subsequent computers have become richer and richer. However, it is interesting to notice that computer scientists are now replicating Turing's original concern by developing new RISC computers (Reduced Instruction Set Computers).

[[7]]   Magnetic tape.

[[8]]   Machine instructions.

[[9]]   For a concrete description of the order codes of the scaled-down computer (Pilot ACE) that was based on this report, see CAMPBELL-KELLY (1982).

[[10]]   The problem of delays actually impinged on the programmers of Pilot ACE where an order code was adopted which required the programmer to incorporate delay information into his program. This cumbersome scheme while producing optimum programs was rather error-prone.

[[11]]   A rather long way of saying that signals will have tolerance ranges.

[[12]]   This makes better sense with a comma after β_2.

[[13]]   The left-hand circuit in Fig. 4 would be better shown as taking its excitory connections from the output shown as the horizontal line.

[[14]]   The reference to "the last figure" concerns the carry digit from the previous addition of the binary digits entering along the input lines.

[[15]] All this and the previous paragraph state that the full power of the computer is realised only if the hardware mechanism of the computer includes facilities for branching.

[[16]] This describes a relatively primitive form of conditional branching. Modern order codes usually provide such a facility as a basic machine instruction.

[[17]] "Subsidiary operation" is a term that was used to describe what we now know as a subroutine.

[[18]] This paragraph describes a stack mechanism for controlling the entry and exit to subroutines. A list of exits—referred to as "notes" by Turing—is required because an operation may enter another operation, which may itself enter another operation, and so on. A single item of storage would not suffice to keep track of this degree of multiple use.

[[19]] While it true that, as Turing states, the mismatch of the mechanical parts of the computer and the electronic parts provides no trouble in transferring data, it does result in a bottleneck. Such a bottleneck was discovered very early in the development of computers; it was one of the main reasons for the development of computer operating systems.

[[20]] This paragraph is a description of a primitive piece of system software known as a bootstrap loader.

[[21]] In this paragraph Turing describes a scheme whereby a symbolic representation can be used for the numerical computer instructions. This helps in debugging a program, as a symbolic representation is much easier to read than a numerical one. Indeed, the second half of the paragraph gives an inkling of what turned out to be a major productivity gain in computing: the provision of symbolic autocodes, and eventually high-level programming languages. Such autocodes consisted of instructions which could be represented symbolically and could be translated, by the computer, into the numerical versions of instructions such as those described by Turing in this paper. These instructions would then be executed normally. For a discussion of autocodes, see CAMPBELL-KELLY (1980).

[[22]] At the time of writing of this paper the British telephone system did not directly use dialing codes for geographical areas but a letter code.

[[23]] Under J.H. Wilkinson the NPL became a world leader in the solution of such problems.

[[24]] This is the first reference to an application area which is only now blossoming.

[[25]] A little-known board game for two or four players. It is played on a board of 256 squares.

[[26]] See *Digital Computers Applied to Games* (this volume).

[[27]] Hardware technology advanced quickly in the years following this paper. Consequently such checks were no longer needed. However, a popular technique used in today's high-reliable systems is n-version programming: a number of versions of a system are developed seperately and are executed in parallel with the results from each execution being checked as being the same.

[28] Presumably, a reference to the checks discussed in the previous paragraph.

[29] This notation is only a way of representing binary patterns.

[30] TS 6 is used as a temporary store, see p. 57.

[31] The continuing presence of P 17 would ensure during the fetch execute cycle that the contents of TS 6 would always be treated as the next instruction.

[32] This is shown in Fig. 19 as the sequence 16, 1, 2, 4, 8 in the CD. The remaining parts of the figure below are the order codes referred to in section 12 of the paper.

[33] This is the description of the proposed fetch-execute cycle of ACE. In essence it differs very little from the fetch-execute cycle of modern computers, although, of course, the details will be different.

[34] Copying into TS 1, 4, 5, or 8 would overwrite important information required for the operation of the computer. The uses of these temporary stores are discussed on p. 57.

[35] Delay line.

[36] There is a contradiction here between this statement and the description of the type B statement towards the top of page 51 where the storage of the old CD is specified as being TS 13.

[37] It is to be presumed that this is a reference to the quantity of input and output of data, since all input and output wil be time-consuming, even compared with the relatively slow cycle time of ACE.

[38] The modern translation of the term *subsidiary* is subroutine.

[39] Strictly, the second column gives the popular form, the first column only acts as a serial numbering convention.

[40] Except of course where the destination is specified in the instruction, as in instruction 11 of INDEXIN.

[41] It is surprising that Turing did not choose a friendlier notation for this, as he had done for other instructions.

[42] See [30].

[43] The use of BURY as instruction 13 is problematic. BURY acts as a subroutine linkage saving mechanism. Its use here would indicate that a further subroutine (subsidiary) is to be called. A better interpretation is that INDEXIN is meant to be used as a subroutine and that an UNBURY is required rather than a BURY. See [45].

[[44]] In the descriptions of DISCRIM, PLUSIND, BURY, UNBURY, MULTIP etc., Turing is setting up a more useful set of instructions than the ones available to him.

[[45]] BURY acts as a link storage mechnism and UNBURY acts as a returning mechanism for subroutine calling. For their operation they require a stack pointer which is contained in TS 31.

[[46]] This should be the minor cycle whose position is given by the contents of TS 31 minus 1 since BURY increments TS 31 by one as its last operation.

[[47]] Before this program is executed the contents of TS 31 should be set to the value of the address where the first of the subroutine links is to be stored.

[[48]] A technique for polynomial evaluation known as Horner's rule.

[[49]] Mercury proved the best medium although delay lines had a short lifetime of popularity.

1947 *Lecture to the London Mathematical Society on 20 February 1947*

[[1]] Typically such analogue machines, which were used extensively until the early seventies, used operational amplifiers and were used to solve systems of differential equations.

[[2]] Developments in hybrid computing: the combination of digital and analogue computers, during the late sixties, enabled systems of partial differential equations to be solved. However, the vast majority of numerical analysis work in industry and academia is carried out using essentially the same machine architecture as described by Turing in this paper.

[[3]] See TURING (1937).

[[4]] Magnetic tapes were the first medium used for large-scale storage of data and are used up to the present day, although they are now being superseded by other media such as floppy discs. However, they were only really suitable for two purposes. First, the storage of data which can be processed serially, for example, in pay-roll calculations when sequences of employee records are read in one record at a time and processed. Second, as a long term archival medium.

[[5]] These problems were solved in the sixties with the use of mechanical magnetic storage devices such as the moving head disc, together with sophisticated indexing schemes.

[[6]] The basis of modern semiconductor random access memories.

[[7]] Turing's original proposal placed before the Executive Committee of NPL envisaged between 50 to 500 mercury tanks, each with a storage capacity of 1000 digits.

[[8]] We now generally regard both memory space and processor speed as equally important. The degree of importance of each really depending on the application which the computer is used on. Turing's remark is probably a reflection of the comparatively immature storage technology of the day as compared with that of electronic circuitry, where the second world war had provided a major impetus.

[[9]] A description of the order code of ACE can be found in CAMPBELL-KELLY (1982) together with a comparison with other order codes used in early and comparable computer systems.

[[10]] This sentence seems to contradict the one following. Presumably what Turing is referring to is the action of the circuit when a code of instruction is recognised and acted upon.

[[11]] Apart from one or two eccentric computers developed in the sixties, the vast majority of computers now use binary notation.

[[12]] This practice was abandoned quite early in computer design. The vast majority of digital computers now have hardware circuits which carry out the operation of division.

[[13]] This is the earliest example of the use of subroutines, or subsidiary tables as they were referred to by Turing. It anticipates the modern view of the architecture of a software system as consisting of chunks of programs (subroutines, modules or procedures) which cooperate with each other by carrying out calculations or some other programming action and passing data to each other.

[[14]] These initial cards are an example of a program known as a loader. Such a program would arrange for the program to be executed to be deposited in suitable memory locations, clear registers and start the program execution process.

[[15]] This is a reference to checks which should be built into a program to ensure that errors known as run-time errors are detected and displayed. A typical run-time error occurs when a programmer writes an instruction which examines a memory location whose number is greater than the maximum number of locations inside the computer. This form of checking is now implemented in the operating system of modern computer, and programmers no longer have to bother with this process.

[[16]] A major concern of numerical analysis, for Turing's contribution see TURING (1948).

[[17]] After a considerable period of little interest this work is now coming to fruition, see WOOFF and HODGKINSON (1987) for an example.

[[18]] This anticipates the explosion of interest in machine learning in the nineteen eighties.

[[19]] Turing's ideas of game-playing by computer have partially been borne out by time. There are a number of computer programs in existence which are capable of beating very good players and one or two which just hover below grandmaster status. However, these programs depend on the massive computational power of the computer rather than on its capacity for learning.

1948 *Intelligent Machinery*

[1] For an amplification and expansion of these objectives, see *Computing Machinery and Intelligence* (this volume).

[2] Writing the sum as

$$(15 + 18 + 21 + \cdots + 54 + \\ 54 + 51 + 48 + \cdots + 15)/2$$

gives this formula.

[3] An example of such a paper game is given in *Digital Computers Applied to Games* (this volume).

[4] A calculating machine of the day.

[5] I.e. algorithmic.

[6] This is just a description of block switching, whereby part of the 9-bit memory address would contain a block number.

[7] Each of the columns is calculated by examining the connections to each unit and applying the multiply and subtract rule.

[8] Routine being used here in the sense of a computing subroutine: a series of pre-programmed instructions.

[9] Turing actually uses A, B, C as externally visible acts in the example below.

[10] A word of explanation is required for the format of this table. The first entry in a row shows how the next situation is determined, what visible action occurs and any changes to the memory or stimulus lines. For example, row 4 shows that when the machine is in situation 4 the next situation is determined by the contents of S1. If S1 is 0, then the next situation is the remainder formed on dividing $2*4$ by 5, if S1 is one, then the next situation is the remainder formed on dividing $2*4+1$ by 5, the next entry in this row shows that visible action A will occur and that memory location M1 is set to 0.

Row 3 shows that the determination of the next situation depends on the substituted value of P (U, T0, D0 or D1) and that visible action B occurs. No memory or stimulus line is set.

[11] Together with *Proposals for Development in the Mathematics Division of an Automatic Computing Engine (ACE)* (this volume) this represents one of the earliest references to the use of subroutines in computer programming.

[[12]] An early prediction of the immense importance of theorem proving in artificial intelligence.

[[13]] See also *Computing Machinery and Intelligence* (this volume) for an expansion on this theme.

1949 *Checking a Large Routine*

[[1]] This paper contains a number of transcriptions errors. These were pointed out in Morris and Jones (1984).

[[2]] The paper was delivered in 1949 at the inaugural conference of the EDSAC computer which had been built at Cambridge. At the time Turing was deputy director of the Manchester prototype computer project.

[[3]] Turing's remarks on the splitting up of a problem into tractable sub-problems foreshadows the sofware engineering use of self-contained chunks of program code known as program units: equivalent to subroutines or procedures, as devices for controlling complexity. Program units can be specified separately and programmed separately by different staff and, as long as the interface between these program units is correctly specified, they can be joined together to form a complete software system.

[[4]] This should be $n!$. This was corrected in Morris and Jones (1984).

[[5]] A transcription error (Morris and Jones 1984). This should be $r!$ and $sr!$.

[[6]] This should read "We can change $sr!$ to $(s+1)r!$ by addition of $r!$" (Morris and Jones 1984).

[[7]] It has been pointed out (Morris and Jones 1984) that the correct way to regard the contents of the boxes in Fig. 1 are not as programming statements, but as specifications which must be satisfied by some programming statements. Thus, box G does not stand for increment variable s by 1, since the box F requires the old value of s to carry out the test $s-r$.

[[8]] This should be "It is also intended that u be $sr!$ or something of the sort e.g. it might be $(s+1)r!$ or $s(r-1)!$ (Morris and Jones 1984).

[[9]] One drawback of the notation adopted by Turing is that it is restrictive: it only allows the explicit expression for the value of each variable of interest, rather than allowing the values of variables to be related to each other. Thus, the inequality $r \leqslant n$ does not appear in Fig. 2, as it has to, in order to infer the $v = n!$ claim at D from $v \geqslant r!$ at C and $r \geqslant n$ by D from C (Morris and Jones 1984).

[[10]] The restrictions on s and r do not, in fact, appear in Fig. 2.

[[11]] This should be condition A (Morris and Jones 1984).

[[12]] This should be $n!$ (Morris and Jones 1984).

[[13]] The maximum storage capacity of the Manchester computer.

⟦14⟧ This should be $u = r!$ and $v' = r!$ (Morris and Jones 1984).

⟦15⟧ This should be $r!$ (Morris and Jones 1984).

⟦16⟧ Transcription error, v should be read as r (Morris and Jones 1984).

⟦17⟧ The first term brackets should be $(n - r - 1)$ (Morris and Jones 1984).

⟦18⟧ The box F should contain the statement TEST $s - r$ (Morris and Jones 1984).

⟦19⟧ The entry for storage location 29 in column E should be n (Morris and Jones 1984).

⟦20⟧ The entry in column F should be WITH $s' = s + 1$. The last line in column A should be $u' = 1$ (Morris and Jones 1984).

⟦21⟧ This paper can be seen as a milestone in the development of the use of mathematics for specifying software and the use of proof methods to check that a particular software system meets its specification. A similar approach to Turing's was developed by Floyd (Floyd 1967) who developed a notation which overcame the problems referred to in ⟦9⟧ and Hoare (Hoare 1969) who formalised the approach. The most sophisticated example of work similar to that reported in this paper is by Dijkstra (Dijkstra 1976). This describes a method for the systematic construction of a program from a pre-condition and a post-condition. The former being a predicate which describes a state involving program variables before a program is executed; the latter being a predicate which describes a state involving program variables after a program has been executed. Although the work reported in this paper is strikingly similar to the research carried out by Floyd, Hoare and Dijkstra, there is no evidence that it directly influenced them (Morris and Jones 1984).

1950 *Computing Machinery and Intelligence*

⟦1⟧ The question of whether A is a man being replaced by the question as to whether A is a machine.

⟦2⟧ An example of this method of construction, by building an adaptive computer is described in *Intelligent Machinery* (this volume).

⟦3⟧ This rather tortuous and cognitively incorrect explanation of man as a computer is an attempt to provide as many intellectual crutches as possible for an audience of readers who would have very little scientific background and, because of the rudimentary nature of computers when this article was published, little knowledge of computers.

⟦4⟧ 17 acts as an instruction identifier. This would now be normally written first.

⟦5⟧ See also *Intelligent Machinery* (this volume).

⟦6⟧ However, the speed of electrical connections in a computer makes it an ideal medium for research into neural processes. For example, see Ru-melhart and McClelland (1986).

⟦7⟧ A reference to the Manchester Prototype, the world's first stored program computer.

⟦8⟧ See also *Intelligent Machinery* (this volume).

⟦9⟧ An interesting version of the Turing test where humans mimicked a computer and where a similar form of deception was practiced is described in Hofstadter (1985).

⟦10⟧ See *Intelligent Machinery* (this volume).

⟦11⟧ A typical differential analyser is the analogue computer popular during the sixties and the seventies. Such computers use voltages to represent physical quantities.

⟦12⟧ For a forceful treatise on the simplicity of man as a behaving system and the consequent application of rules to describe behavior, see Simon (1981).

⟦13⟧ This form of programming is now a very active research area, after a brief vogue in the sixties (Rumelhart and McClelland 1986).

⟦14⟧ A detailed example of this is given in *Intelligent Machinery* (this volume).

1953 *Digital Computers Applied to Games*

⟦1⟧ A statement of one the major problems in computerised game playing; particularly in chess, where top-class chess players tend to decide on moves on the basis of pattern matching as well as the computational procedures described in this paper. See de Groot (1966).

⟦2⟧ Chess has five key features which make it an excellent medium to explore issues about human knowledge, and the transfer of such knowledge to computers. These are: it is a fully defined and well-formalised domain; it offers a challenge to the highest levels of human intellectual capacity; it involves a large range of issues in knowledge representation and cognitive functions such as: logical calculation, rote learning, concept formulation and inductive reasoning; a large degree of knowledge about the game has been accumulated; and an accurate and generally accepted scale of performance is available (Michie 1982). In general chess offers a laboratory for artificial intelligence workers which enables them to explore ideas which can be employed in more utilitarian applications such as intelligent tutoring, disease diagnosis and electronic circuit fault-finding.

⟦3⟧ Progress in using computers for predicting horse races or football matches has not progressed very much since the time when this article was written. See Drapkin and Forsyth (1987).

⟦4⟧ Now a trivial programming task; one which would not even be posed as an undergraduate project.

⟦5⟧ A problem which has now been solved.

⟦6⟧ The answer to this question now is yes. Computer chess programs, such as the CHESS series from Northwestern University, have been developed which are capable of beating the vast majority of chess players, and approach grand-master status.

⟦7⟧ Still an open research question. For an impressive example of the use of machine learning applied to chess, see SHAPIRO (1987). However, even the most vigorous proponents of machine learning would agree that a considerable amount of research into artificial intelligence is required before computer programs, based on machine learning, achieve the performance of current chess programs based on a computational approach.

⟦8⟧ The answer to this question is, almost certainly, yes. Progress in computer chess has been such that chess playing programs are now available which are considerably better than the majority of chess players.

⟦9⟧ This is still an open research question.

⟦10⟧ The reason for this rule being inappropriate is the combinational explosion of possible moves that would occur in chess where, in comparison with noughts-and-crosses, the number of future moves is much smaller.

⟦11⟧ This is one of the earliest references to the idea of an evaluation function. A concept used extensively in artificial intelligence; in particular, in game playing.

⟦12⟧ This is one of the earliest references to tree pruning. This is a technique which is extensively used in computer game playing, which enables a game playing program to cut down the width of its search space. Much more sophisticated tree pruning methods are used in modern programs.

⟦13⟧ This is an example of the detailed calculation of an evaluation function. Current evaluation functions would be calculated from both the value of position totals and the position play value. The original invention of this function is credited to Claude Shannon (SHANNON 1949).

⟦14⟧ This remark is borne out by the fact that, currently, one of the most active research areas in computer chess involves the use of knowledge bases for end-game processing, rather than the computational approaches, described in this article, which tend to be highly effective for the middle game.

⟦15⟧ A statement which is now regarded as clearly wrong. Many research workers in computer chess, while often being average to good chess players, have built programs which are far in excess of their playing abilities.

⟦16⟧ There have been no succesful attempts to carry out this learning process in computer chess. The nearest that the developers of automated chess playing programs have reached is the manual adjustment of the evaluation function in order to achieve minor gains in performance. However, a highly succesful draughts playing program (SAMUEL 1959) is partly based on an algorithm that adjusts the evaluation function.

[17] Considering the primitiveness of hardware and software facilities in the early fifties this program must be counted as one of the great software achievements.

[18] For example by keeping areas of the chess board which are densely populated by pieces in the main memory of the computer.

[19] And, consequently, was the first area where highly skillful programs were developed (SAMUEL 1959).

[20] The representation of the draughts board is chosen in order to minimise storage space and programming complexity. An alternative representation—and perhaps a more natural one—would be to allocate three bits for each square. The number 0 would represent an empty square, the number 1 a black piece, the number 2 a white piece, the number 3 black king and the number 4 a white king. This representation would occupy the same space as that chosen by Turing. However, the programming involved would be more complex.

[21] Also known as logical and.

[22] This description of the moves provides the rationale for the numbering convention used on the draughts board. By using this convention, fast multiplications by 2 can be used in calculating moves and checking board occupancy.

[23] The symbol $\equiv$ stands for logical equivalence.

[24] A modern draughts playing program would not fall into this trap because advances in tree pruning and hardware technology would lead to a deeper search. The solution adopted by Turing to overcome this problem can be seen as a rather ad-hoc correction.

[25] See [24].

[26] Since k is equal to 1.

1954 *Solvable and Unsolvable Problems*

[1] Simply stated the rule is that the transformation is possible if an even number of interchanges leads to the required final state of the puzzle.

[2] One of the earliest references to the partitioning of problems into those that are undecidable and those that are intractable. While the statement that the investigations which are not concerned with the amount of work involved are easier to carry out, was certainly true at the time of writing, there is now a large corpus of work on computational complexity and, consequently, this gap has been narrowed.

[[3]] The term *not perfectly regular* means that, for example, one card may not completely cover another or may not symmetrically overlap another card.

[[4]] Where the initial state of the puzzle would be some set of premises and axioms, the final state the required theorems and the transformation mechanism being provided by laws of inference.

[[5]] It is interesting that Turing dismisses a question which is so vitally important to computer scientists, and on which there is now a considerable corpus of theoretical work. This is especially surprising considering the fact that computer hardware was sufficiently undeveloped in the 1950s that there was a major onus on programmers to optimise their use of computer time.

[[6]] This choice of $P(R, R)$ might seem an odd one. However, it does set up the conditions for a particularly elegant demonstration on p. 19.

[[7]] p. 17.

[[8]] See GÖDEL (1931).

[[9]] See NOVIKOV (1952).

BIBLIOGRAPHY

CAMPBELL-KELLY, M.
1980   Programming the Mark 1: Early programming activity at the University of Manchester
        Ann. Hist. Comput. **2** (2), 130–168
1981   Programming the Pilot Ace: Early programming activity at the National Physical Laboratory
        Ann. Hist. Comput. **3** (2), 133–162
1982   The development of computer programming in Britain (1945–1955)
        Ann. Hist. Comput. **4** (2), 121–139

CARPENTER, B.E. and R.W. DORAN
1977   The other Turing machine
        Comput. J. **20** (3), 269–279

COLBY, K.M.
1963   Computer simulation of a neurotic process
        In: S.S. TOMKINS and S. MESSICK (Eds.), *Computer Simulation of Personality: Frontiers of Psychological Research*
        (Wiley, New York) 165–180
1964   Experimental treatment of neurotic computer programs
        Arch. Gen. Psychiatry. **10**, 220–227

DE GROOT, A.D.
1966   Perception and memory versus thought: Some old ideas and recent findings
        In: B. Kleinmuntz (Ed.), *Problem Solving*
        (Wiley, New York) 19–50

DENNETT, D.C.
1978   *Brainstorms: Philosophical Essays on Mind and Psychology*
        (MIT Press, Cambridge, MA)

DIJKSTRA, E.W.
1976   *A Discipline of Programming*
        (Prentice-Hall, Englewood Cliffs, NJ)

DORAN, R.W.
        see CARPENTER, B.E.

DRAPKIN, A. AND R. FORSYTH
1987   *The Punter's Revenge: Computers in the World of Gambling*
        (Chapman and Hall, London)

DREYFUS, H.L.
1972   *What Computers Can't Do: A Critique of Artificial Reason*
        (Harper and Row, New York)

FLOYD, R.W.
1967   Assigning meanings to programs
        Proc. Sympos. Appl. Math. **19**

FORSYTH, R.
        see DRAPKIN, A.

GALLIER, J.H.
1986   *Logic for Computer Science*
        (Harper and Row, New York)

GÖDEL, K.
1931   Über formal unentscheidbare satze der Principia Mathematica und
        verwandter Systeme, I
        Monatsh. Math. Phys. **38**, 173–198

GOOD, D.I.
1985   Mechanical proofs about computer programs
        In: C.A.R. HOARE and J.C. SHEPHERDSON (Eds.), *Mathematical
        Logic and Programming Languages*
        (Prentice-Hall, London)

HAREL, D.
1987   *Algorithmics—The Spirit of Computing*
        (Addison-Wesley, Reading, MA)

HAYES, I. (Ed.)
1987   *Specification Case Studies*
        (Prentice-Hall, Englewood Cliffs, NJ)

HOARE, C.A.R.
1969   An axiomatic basis for computer programming
        Comm. ACM **12** (10), 576–580

HODGES, A.
1983   *Alan Turing—The Enigma of Intelligence*
        (Hutchinson, London)

HODGKINSON, D.
        see WOOFF, C.

HOFSTADTER, D.R.
1985   A coffeehouse conversation on the Turing test
        In: *Metamagical Themas: Questing for the Essence of Mind and Pattern*
        (Viking, New York)

HUSKEY, H.D.
1984   From ACE to G-15
        Ann. Hist. Comput. **6** (4), 350–371

JONES, C.B.
        see MORRIS, L.

McCLELLAND, J.L.
        see RUMELHART, D.E.

MICHIE, D.
1982   Computer chess and the humanisation of technology
        Nature **299**, 391–394

MORRIS, L. AND C.B. JONES
1984   An early program proof by Alan Turing
        Ann. Hist. Comput. **6** (2), 129–143

NOVIKOV, P.S.
1952   On algorithmic undecidability of the problem of identity
        Dokl. Akad. Nauk. SSSR. **85**, 709–712

PARTRIDGE, D.
1987   To add AI, or not to add AI
        In: B. KELLY AND A. RECTOR (Eds.), *Research and Development in Expert Systems V*
        (Cambridge University Press, Cambridge) 3–13

RUMELHART, D.E. AND J.L. McCLELLAND
1986   *Parallel Distributed Computing*
        (MIT Press, Cambridge, MA)

SAMUEL, A.L.
1959   Some studies in machine learning using the game of checkers
        IBM J. Res. Develop. **3**, 211–229

SEARLE, J.
1980   Minds, brains and programs
        Behavioral Brain Sci. **3**, 417–457
1982   The myth of the computer
        New York Rev. Books, April 29, 3–6

SHANNON, C.
1949   Programming a computer for playing chess
        Philos. Mag. **41** (7th series), 265–275

SHAPIRO, A.D.
1987   *Structured Induction in Expert Systems*
        (Turing Institute Press, Edinburgh)

SIMON, H.A.
1981   *The Sciences of the Artificial*
        (MIT Press, Cambridge, MA)

TURING, A.
1937   On computable numbers, with an application to the Entschei-
        dungsproblem
        Proc. London Math. Soc. (2) **42**, 230–265
1948   Rounding-off errors in matrix processes
        Quart. J. Mech. Appl. Math. **1**, 287–308

WEIZENBAUM, J.
1966   ELIZA—A computer program for the study of natural language
        communication between man and machine
        Comm. ACM **9**, 36–45

WINSTON, P.H.
1979   *Artificial Intelligence*
        (Addison-Wesley, Reading, MA)

WOOFF, C. AND D. HODGKINSON
1987   *muMATH: a Microcomputer Algebra System*
(Academic Press, New York)

.

INDEX

Printed and bound by CPI Group (UK) Ltd, Croydon, CR0 4YY

03/10/2024

01040330-0015